LEGERDEMAIN

Deceit, Misdirection and Political Sleight of Hand
in the Disappearance of Amelia Earhart

By
David K. Bowman

authorHOUSE

1663 Liberty Drive, Suite 200
Bloomington, Indiana 47403
(800) 839-8640
www.AuthorHouse.com

© 2005 David K. Bowman. All Rights Reserved.

No part of this book may be reproduced, stored in a retrieval system, or transmitted by any means without the written permission of the author.

First published by AuthorHouse 05/06/05

ISBN: 1-4208-2983-1 (sc)
ISBN: 1-4208-2984-X (dj)

Library of Congress Control Number: 2005901638

Printed in the United States of America
Bloomington, Indiana

This book is printed on acid-free paper.

Cover photo credits:

Upper left cover photo of Amelia Earhart from author's collection

Upper right cover photo of Irene Bolam, taken in Yugoslavia, 1976, Courtesy of Diana Dawes

Cover photo overlay of Amelia Earhart/Irene Bolam, courtesy of forensic study by Tod Swindell copyright 2002.

Other Photo Credits

Photo of Irene Bolam used in Dave Deal's overlay on page 298, provided through the gracious permission of the late Joseph Gervais.

DEDICATION

To the Earhart researchers everywhere who have never given up . . . and of course to Amelia Earhart, wherever she is.

legerdemain 1. Slight of hand. 2. Any deception or trickery; hocus pocus. [Middle English *legerdemayn,* from Old French *leger de main,* "light of hand"; *leger,* light, from Vulgar Latin *leviarius* (unattested), from Latin *levis,* + *main,* hand, from Latin *manus.*

Table of Contents

Illustration List xi
Acknowledgements xiii
Foreword xv
Introduction xvii

A Rising Star 1
The Official Story 4
Islands of Mystery 7
Busy 1960, Fred Goerner And Admiral Nimitz 13
Crash Near Bicycle Lake 18
Operation Earhart 20
The Bolam Affair 23
The Enigma of William Van Dusen 28
The Bizarre Tale of Wilbur Rothar 33
Electra: The Sleight of Hand Begins 40
More Sleight of Hand: Secret Modifications 47
The Route 49
The ITASCA 53
Miss Earhart Feared Forced Down at Sea 60
Interception 67
What will they do to us? 70
The British Connection 80
Lost Star 83
One Night on Saipan 88
An Unclaimed Letter in Jaluit 99
Notes in a Bottle 102
The Ghosts of Garapan Prison 106
The Morgenthau Transcript 127
The Eyewitnesses 133
Other Strange Reports from World War II 137
Case Closed? 141
Deep Sea Searches 144
Tighar 147
Rollin Reineck, Forensics, and a Second Mrs. Bolam 149
Tokyo Rose 155
The Carrington Report 158
The FBI File 160
We Must Be On You But Cannot See You 164

Secret Cruises 169
The Putnam Papers 172
What Lies Beneath 174
Dark Agenda 178
The Real McCoy? 182
Legerdemain: Putting Together What Really Happened 185
What Finally Happened to Earhart and Noonan? 192
Echoes of a Cover-up 197

Afterword 207

Appendix I - Timeline 210
Appendix II - CAA Numbers on Amelia Earhart's Electra 258
Appendix III - Documents Relating to Amelia Earhart's Electra 260
Appendix IV - Text of 1939 Gendarmerie Report 265
Appendix V - Amelia Earhart Flight Covers 268
Appendix VI - History of Paul Mantz' Electra 274
Appendix VII - ITASCA Radio Logs 277
Appendix VIII - Morgenthau Transcript and Related Documents 286
Appendix IX - Pros and Cons - Was Irene Bolam Amelia Earhart? 298
Appendix X - History of Howland Island 301
Appendix XI - Selected Website List 303
Appendix XII – Photo Overlays by David Deal 305
Bibliography 310
Index 315

Illustration List

"Large Throng Welcomes Amelia" 6-20-32 112
Hubby Proud of "Lady Lindy" (Feb. 1931) 113
In the Spotlight of the World (1932 article) 114
President Honors Miss Earhart (1932 article) 115
Midweek Pictorial cover, 1935 116
Enlargement of a photo of "Wilbur Rothar" 117
Miscellaneous Clippings from the Rothar Affair 118
USCGC "Summary of Search" (Page 104) 120
USCGC "Summary of Search" (Page 105) 121
"Where Fate of Flyers is Hidden in Pacific" 7-9-37 122
"Putnam is [sic] Kidnaped" 5-13-39 123
Amelia Earhart Still Alive Believes Mother (4-18-49) 126
Earhart souvenir cover order form – outside view 271
Earhart souvenir cover order form – inside view 271
Round the World Cover (First Attempt) – Unused 272
Round the World Cover (First Attempt) – Cancelled 272
ITASCA commemorative cover (Bellarts copy) 273
Title page of National Archives ITASCA radio log microfilm 278
Original Bellarts ITASCA radio log 1-2 July 1937 (1st Page) 279
Original Bellarts ITASCA radio log 1-2 July 1937 (2nd Page) 280
Original Bellarts ITASCA radio log 1-2 July 1937 (3rd Page) 281
National Archives copy ITASCA radio log 1-2 July 1937 (1st Page) 282
National Archives copy ITASCA radio log 1-2 July 1937 (1st Page) 283
National Archives copy ITASCA radio log 1-2 July 1937 (3rd Page) 284
Howland Island radio log RM2 Cipriani 2 July 1937
(National Archives) 285
Letter from Paul Mantz to Eleanor Roosevelt of April 26, 1938, Pg. 1 ...286
Letter from Paul Mantz to Eleanor Roosevelt of April 26, 1938, Pg. 2 ...287
Administrative note dated 6-7-38 288
Note to Henry Morgenthau from Eleanor Roosevelt of May 10, 1938 289
Morgenthau Transcript, telephone conversation May 13, 1938 Pg. 411 ...290
Morgenthau Transcript, telephone conversation May 13, 1938 Pg. 412 ..291
Morgenthau Transcript, telephone conversation May 13, 1938 Pg. 413 ..292
Carbon of letter to Paul Mantz from Eleanor Roosevelt May 14, 1938 ...293
Undated note on White House stationery 294
Note to Eleanor Roosevelt from Morgenthau 7-5-38 295
Note from Eleanor Roosevelt to Morgenthau 7-5-38 296
Letter from RADM Wesche to Paul Mantz of July 21, 1938 297

David Deal Overlay: Amelia Earhart to Irene Bolam 307
David Deal Overlay: Irene Bolam to Amelia Earhart 308
David Deal Overlay: Amelia Earhart to Irene Bolam 309
Literary Digest, 6-30-28 111
Search Pattern of USCGC ITASCA 2-18- July 1937 119
1936 government photo of a Lockheed Electra Model 12A 124
Lobby card from the film "Flight for Freedom" 125

Acknowledgements

I'd like to first thank David Bellarts for his assistance with this book. He made available many items, including some of his father's correspondence with various people relating to Amelia Earhart, as well as a copy of Chief Bellarts' radio log from Howland Island. In more ways than one, David's assistance was pivotal to the creation of this book.

Next, I'd like to warmly thank Tod Swindell for graciously giving permission to use one of his photographic overlays on the dust jacket of this volume. It is the ideal graphic for LEGERDEMAIN and is deeply appreciated.

Additional warm thanks go to Ann Pellegreno for her gracious permission to use the photo of Irene Bolam on the front cover this book. She is a great aviator, whom I am proud to call a friend.

I'd also like to thank Ron Reuther, original moderator of the Amelia Earhart Discussion Group on Yahoo for his gracious assistance with my book. He is a researcher extraordinaire. His timeline, included as Appendix I, is exceptionally informative and indispensable for the Earhart researcher.

Dr. Alex Mandel, Ph.D., a member of the Earhart Discussion Group, and authority on naval history, was exceptionally helpful to me and was indispensable in helping me with a final review of the manuscript. He helped me catch many errors and discrepancies. Thank you, Alex.

Also, my thanks go to noted Amelia Earhart researcher Ron Bright, who provided me with some important information, which I really appreciated.

I must also thank David Deal, who supplied the provocative and interesting photographic overlays shown in Appendix XII. Never before published, they were added to this book just before it went to press. Their use is deeply appreciated. Thanks, Dave.

Also to be thanked are my brother, Thomas Bowman, and his wife, Hong, who provided invaluable assistance in manuscript preparation.

My thanks also go to all of the other members of the Amelia Earhart Discussion Group, many of them noted Earhart researchers and writers, including the current moderator, Michelle Cervone, for their warmth and gracious assistance.

Additionally, I must also thank TIGHAR (The International Group for Historic Aircraft Recovery) for their assistance. They have a fascinating website, which is a wonderful source of information for the researcher.

In the writing of this book, I relied heavily on the work of numerous other researchers and writers, and would like to gratefully thank each and every one of them. I would especially like to thank Joe Gervais, Joe Klaas, Robert Dinger, Paul Briand, Don Wilson, Randall Brink, Robert Myers, Frederick Goerner, John P. Riley, Buddy Brennan, Thomas Devine, and James Donahue, the last six posthumously.

Every researcher developed and presented some sort of important and valuable information to contribute to the body of knowledge regarding the disappearance of Amelia Earhart. And thanks to their efforts, the world is closer to knowing the complete truth regarding the disappearance of an amazing human being.

Writing this book was a deeply satisfying experience. To me, it was the most rewarding and gratifying way one could start a retirement. I hope that this work serves a constructive purpose, and is appreciated by my readers.

David K. Bowman
Buckley, WA
January 2005

Foreword

Before beginning this book, the reader should know that the author does not endorse every facet of every account presented in this book. With all due respect to the researchers whose work I discussed, from a logical standpoint, alone, it would be impossible.

Aside from presenting the results of his own investigation, it is the author's intent to review and correlate, in an interesting manner, all the significant events relating to Amelia Earhart's disappearance, and all the significant investigations that followed. Ultimately, it will be for the reader to decide the answer. A major thing that continues to attract readers and writers alike to the disappearance of Amelia Earhart is its core of ultimate mystery, an impenetrable "heart of darkness", if the reader will. For two-thirds of a century, many have tried to pierce this heart of darkness to discover its secret.

Undoubtedly, another thing that keeps attracting readers and writers alike to this subject, aside from the unsolved dramatic disappearance of a highly charismatic personality, is the romance of global aviation in an era that is gone forever. Even the area in which Earhart went missing, one of the most exotic areas of the world, was all the more exotic, mysterious and dangerous in the 1930s. It was a place rife with suspicion, espionage, and treachery, a place where you had to watch your back.

Finally, it has been said that "getting there is half the fun". The biggest aim of this book is to not only try to solve a mystery, but to enjoy, along the way, an amazing story of adventure, mystery and intrigue in a long vanished, art deco world.

Introduction

On the sultry morning of July 2 1937, a gleaming new Lockheed Electra just barely lifted off the ground near the end of the runway at Lae, New Guinea. It winged uncertainly toward the morning sun. Where that plane actually flew and what actually happened aboard it during the next twenty-four hours remains a matter of controversy to this day.

The strange part of this story, the story of the disappearance of Amelia Earhart, is that the truth has been there all along, in the memories and recollections of those who were involved, in a few private files, and in the closely guarded classified files of the U.S. Government.

Although the disappearance of Amelia Earhart is always cited as one of the greatest aviation mysteries of the 20th Century, it has never, for the most part, been a real mystery. Not to anyone in the right governmental or social circle, not to anyone in the know. It has only been a mystery to the public at large. This is one of the things that this book will show.

This book came about after I retired a short while ago. I began to read more about the Earhart disappearance and was struck by the fact that there has been little correlation of the various published investigations over the years. As a result, many of the investigations, have gone waffling off into their own particular avenue of research, evidently unaware that some new information turned up by another investigation completely moots their efforts.

And, even more comically, some investigators seem to have picked up on the new data, but ignored it anyway, as it does not fit their thesis.

As a result, I saw a need for some sort of comprehensive correlation of material developed to date, so that the reader can read of the various aspects of the disappearance of Amelia Earhart in a single volume.

Moreover, I saw a need for some new checking on my part, using some of my newfound leisure to more fully investigate the Earhart disappearance.

The final aim of this book has been to try to come up with an unbiased and factual explanation for one of the most controversial aviation mysteries of the 20th century.

The reader will note that I have not "hard sold" any particular theory. I have just tried to piece the facts together and go where they take me. However, it should be borne in mind that the material presented in the summary chapters is merely a possible solution to a most perplexing mystery.

What emerged from my research was what the title of this book conjures up: a picture of misdirection, deceit, switched planes, clandestine aircraft modifications, secret flight routes, false news releases, and just maybe . . . cold blooded treachery.

I found in my research legerdemain of the highest order, on the part of both the U.S. and Japanese governments. And while it must be said that a certain amount of the U. S. Government's misdirection was probably perpetrated for patriotic purposes to support an important intelligence-gathering mission, some of it was purely to cover up ineptitude, incompetence, bad planning and God knew what else. Moreover, one of the strangest things about the whole affair is that two-thirds of a century later it is still under wraps officially.

Also, along the way in my research, I developed some new insights and perspectives, which will hopefully add to the body of information on the Earhart disappearance.

And now, enter the world of Legerdemain.

Chapter 1

A Rising Star

On a stormy day in June 1928, the tri-motor seaplane Friendship dipped down out of a leaden sky to land in the bay near Burry Port, Wales. The airship taxied through the pouring rain to a nearby buoy and cut its engines. A moment later, one of the crewmen, Louis "Slim" Gordon, opened a door in the fuselage, hopped out, and moored the airship to the buoy.

At the controls of the airship was Wilmer Stultz, and in the passenger compartment was a woman, who up until that flight, had been a recreational aviator and a social worker in Boston. On that rainy morning, she was catapulted to international celebrity. She was Amelia Earhart.

A few months before, the young, boyish woman with tousled hair, had been asked to an interview by George Palmer Putnam, the wealthy and powerful head of G.P. Putnam's Sons, publishers. An athletic adventurer, writer and promoter, Putnam had been asked by a wealthy New England socialite to find a woman to fly across the Atlantic in the aircraft she had purchased from Admiral Richard Byrd. She had initially planned to make the flight herself, to be the first woman to fly across the Atlantic, but had been pressured by her worried family out of the flight.

A masterful promoter, George Palmer Putnam, or "GP" as he liked to be called, immediately seized upon the young Amelia Earhart at their first meeting. Earhart was tall, slim and had a remarkable physical likeness to recent aviation hero Charles "Lucky Lindy" Lindbergh. Putnam instantly christened her "Lady Lindy", a nickname which Earhart deplored.

When asked if she wanted to make the flight, Earhart unhesitatingly jumped for it, as she was plucky, adventurous, and ambitious. She knew an opportunity when she saw one. As she said later, "You don't turn down an opportunity like that!"

The rest was history. Immediately after the flight of the Friendship, commemorative medals were struck and sold, and the young aviatrix embarked upon a number of product endorsements.

At the same time, GP hustled his young protégée off to his luxurious estate, Rocknoll, in Rye, NY, so that she would have the privacy and peace to write an account of her famous flight. This she did, and before the end of 1928, "20 Hours, 40 Minutes" was published. To this day, it is an important historical source and a sought after collectible.

Earhart then embarked upon a lucrative and busy lecture tour to discuss her new book. Upon her return to New York, she was appointed Aviation Editor for Cosmopolitan Magazine. By then, in 1929, she was all the rage.

Dissatisfied with being just a passenger on the first Transatlantic flight, Earhart determined to pilot the Atlantic Ocean herself and spent the next four years in preparing for this. Preparations were set back by a crash during a practice flight in Norfolk, Virginia in 1930, which necessitated lengthy repairs that weren't completed until 1931. Shortly after the Norfolk crash, GP obtained a divorce from his wife, Dorothy, and the following February, in 1931, he and Earhart were married in a quiet ceremony.

In the spring of 1932, the aviatrix took off from Newfoundland and successfully crossed the Atlantic in 15 hours, 18 minutes, landing in a pasture in Londonderry, Ireland. Earhart was now the first woman to successfully pilot the Atlantic.

Over the next five years, under GP's guidance, Earhart set more aviation records, participated in various aviation events, continued to tour the lecture circuit, was the spokesperson for a multitude of products, and lent her name to several businesses. One of them was a line of women's clothing, which she personally designed. Another was a high quality line of luggage that continued be manufactured for years after her disappearance.

Additionally, Earhart became actively involved in establishing commercial air routes and founding airlines. But beyond that, Earhart was an ardent feminist, who eschewed the conventional female role and forged a new one for herself. On her passport, under "occupation", she had entered, "flyer". And whenever she was on an airfield, she habitually wore custom tailored gabardine slacks, an open throated man's sport shirt with a knotted silk scarf, and a leather-flying jacket. They became her trademark. She was one of the most talked-about, fashionable, admired, beloved, and emulated women of the 1930s. She was an icon. Her name was a household word. Even the press referred to her more often as not as just "Amelia." Everyone knew who they were talking about.

During her brief career, she was always thinking about the next flight, as it was her flights that kept her in the public eye and kept her career going. During her preparations for the round-the-world flight, she told a friend, "I think I've only got one more good flight left in me". That remark turned out to be more prescient than Earhart could know.

Chapter 2

The Official Story

On the humid morning of March 20, 1937, after over a year of preparation, Amelia Earhart attempted to take off on the second leg of her round-the-world flight from Luke Field, Hawaii, heading west. Her next stop was to be Howland Island, in the Central Pacific.

Something went very wrong during Earhart's takeoff run. Just seconds before the big plane would have been in the air, it began to fishtail wildly from side to side in what aviators call a ground loop. Earhart had momentarily lost control. Moments later, the landing gear failed under the huge stresses involved, and the machine violently collapsed onto the runway and slid to a stop with a shriek of rending metal. Although witnesses recounted that Earhart had coolly handled the situation, nevertheless, it was a shaken aviatrix who climbed ashen faced out of the wrecked aircraft.

"I don't know what happened, Paul," Earhart said to Paul Mantz, her technical advisor, and one of the first to arrive on the scene.

"That's alright," said Mantz, as he wrapped a comforting arm around Amelia's shoulders. "As long as nobody was hurt. You just didn't listen to papa, did you?" This last remark, from Paul Mantz, was a telling one regarding the relationship between the two.

Shortly afterward, news releases proclaimed that the damaged airplane was being dismantled and shipped to the Lockheed plant in Burbank, California for repair.

Two and a half months later, on June 1, 1937, amid a splash of publicity, Amelia Earhart and her navigator Frederick Noonan, took off from Miami, headed south, on her second attempt at a round-the-world flight. Just prior to their takeoff, Earhart reportedly left behind her trailing antenna, telegraph key and its accessories.

LEGERDEMAIN

Over the next month, a steady stream of news releases and photos poured back to the U.S., feeding the endless appetite for news of Earhart's world flight. The flight seemed to go off smoothly with few problems, due mainly to GP's (George Palmer Putnam's) organizational skills. However, before Earhart landed in Lae, New Guinea, things started to go amiss. Her Western Electric radio began to malfunction; repairs would have to be sought at Lae. Then the weather soured, delaying their takeoff. Also, there were concerns over weight and whether some items of equipment should be left behind. On the relatively short, dangerous runway at Lae, takeoff weight was the critical difference between success and failure.

Another little known edge that Earhart had was the new technology of carrying two different types of fuel, low-octane, and high-octane. During takeoff, particularly at airports like the one at Lae, she used the high-octane fuel for added boost. Once airborne, she would switch to the low-octane fuel for normal cruising.

The night before they finally did take off, Fred Noonan reportedly went out with a local employee of New Guinea Airways, Earhart's host organization, and partied until about 7:00 a.m. the next morning. Half an hour later, Earhart would knock on his hotel door to rouse him.

A communication from Earhart to GP the day before indicated a problem with "personnel unfitness". But to this day, it is not clear whether Earhart was referring to Noonan's reported propensity to drink or to the dysentery which she had contracted earlier in the flight. Noonan, however, was known by former associates to be used to his heavy-living lifestyle and always arrived on time for flights, performed at a high level. Therefore it is difficult to say whether Noonan was really inebriated at the time of takeoff from Lae.

To be sure, the takeoff on the morning of July 2, 1937 was less than "textbook". Even with the added boost of the high-octane fuel, it was rough. At the end of the runway, the aircraft joltingly lost altitude for a moment, it's landing gear hitting the ground and leaving a cloud of dust. The Electra continued toward the sea, losing altitude until its rear landing gear briefly touched the sea and left a plume of water. It then began to gradually gain altitude at last. There is a high likelihood that the navigator was knocked around as the big aircraft violently lost altitude at the end of the airstrip, although we may never know for sure. Most of the problem had been caused by the weight of the extra fuel that they were forced to carry aboard the plane for the long and dangerous trek to Howland Island.

David K. Bowman

What happened between then, 10:00 a.m. July 2, 1937 and the next morning, also July 2, 1937 due to the International Dateline, is not known for sure. After an occasionally cryptic series of message interchanges between Earhart and the ITASCA, during which neither party ever talked directly to the other, the airwaves fell silent and nothing more was heard. The most charismatic aviatrix of the 20th Century had slipped enigmatically into history. Her time had passed.

Chapter 3

Islands of Mystery

A central factor in the disappearance of Amelia Earhart was the political situation at the time of her round-the-world flight. At the time, Franklin Roosevelt, or "FDR" as he was known, was in the oval office. The world of the mid 1930s was filled with tension, suspicion, aggression, war.

Having served as Assistant Secretary of the Navy for the Office of Naval Intelligence during WWI, FDR came to the presidency with unprecedented experience in and fascination with intelligence.

Coupled with his fascination with intelligence, Roosevelt had a great propensity for secrecy, which has only become apparent in recent decades. For FDR, intelligence blended with politics into a subtle and ruthless game, which he played with consummate skill.

Many of Roosevelt's secret activities turned out to be well taken and well advised, but some of them were less forthcoming in intent, having the flavor of almost a secret international chess game. Much may never be known of FDR's activities, as he had a standing rule of never putting a lot of things in writing, thus keeping them verbal. This was in marked contrast to the policy of later presidents such as Kennedy, Johnson, and Nixon, who tried to tape record every word uttered in the oval office for posterity.

One of FDR's more well-taken procedures was to route his outgoing message traffic through U.S. Army communications, and incoming messages through U.S. Navy facilities. As a result, only the files in FDR's "Map Room" were complete, and nobody in either armed service had a complete overview of his message activity.

Until 1941, when the Office of Information Coordination was established, intelligence collection methods of the government were not always very organized. The OIC would become the Office of Strategic Services (OSS), which would eventually become the CIA in 1947.

FDR was noted during the 1930s for asking wealthy and connected friends and amateurs to accept intelligence assignments during their travels. For some reason, although FDR had at his disposal excellent signal intelligence (intercepted radio messages, etc.), he placed a far greater confidence in intelligence gathered through human means by spies.

The epitome of FDR's fixation on human intelligence was a group called "The Room". It was an intelligence gathering group of wealthy, highly placed people in FDR's circle, which was organized and run by publisher Vincent Astor. FDR's dream was of establishing a vast elaborate network of intelligence operatives to penetrate any government, any facility.

Astor had created "The Room" in 1927 as a purely recreational exercise in low level intelligence gathering with a number of wealthy friends. Astor's fellow members included, among others, Theodore Roosevelt's brother Kermit; banker Winthrop W. Aldrich; diplomat David E. Bruce; publisher Nelson Doubleday and philanthropist Rhinelander Stewart.

The group met monthly in an innocuous apartment at 34 East 62nd Street in New York City, which was equipped with a mail drop and an unlisted phone. They swapped gossip and low-grade intelligence, as well as occasionally entertaining speakers such as Somerset Maugham and Commander Richard E. Byrd. It was a close and select group indeed.

After Roosevelt ascended to the presidency, Astor's quiet informal network of globetrotting wealthy friends became an important and regular intelligence source for FDR.

In the tense world of the 1930s, the government needed all the intelligence it could get to deal with an impending war. Seen in that light, it would be almost inconceivable that an around the world flight proposed by an internationally famous and beloved aviatrix would fail to be seen as the ideal vehicle for some sort of intelligence gathering.

As FDR sometimes played things close and tight, according to a crisp set of rules, including "if you're caught, you're on your own," Earhart's flight would be an eerie precursor of the U-2 affair of 1960. That incident involved military pilot Gary Francis Powers, who flew U-2 surveillance planes for the CIA. In 1960, Powers' plane was shot down and he was taken prisoner. He went through a harrowing period of captivity, before the Russians finally released him. But initially, the U.S. Government disowned Powers' surveillance flight. It was only massive publicity which forced the issue into the open.

In the world at large in the 1930s, there was deep trouble in all directions. In Europe, Adolf Hitler and his National Socialist Party were on the rise. In 1933, the National Socialist Party gained control of the Reichstag, electing Hitler as the head of the government. Almost immediately, Hitler initiated his grisly holocaust and made plans to begin annexing Europe.

In 1935, Italy invaded Ethiopia, which provoked deep concern on the part of the various western allies, who began to fear additional Italian aggression would occur at any time along the Mediterranean.

In the Pacific Area, things were also becoming serious. Back in 1919, after WWI, Japan had been mandated the Caroline, Marshall, and Marianas Islands by the League of Nations. However, there was a big limitation placed on the mandate by the League of Nations, and that was that under no circumstances must Japan militarize any of the islands placed under its trust.

This caveat Japan soon violated by beginning construction of military facilities on some of the islands. From then on until Amelia Earhart's last flight in 1937, anyone unlucky enough to be caught by the Japanese in their mandated territory, without a good explanation for their presence, was at high risk for execution. Unfortunately, this fate befell some of them.

Japan invaded Manchuria in 1931, causing a major outcry among many nations, and escalating international tensions. In 1932, Japanese forces shelled Shanghai in order to quell growing anti-Japanese sentiment. Diplomatic protests and pressure continued, finally reaching the point in 1933 that Japan withdrew from the League of Nations.

As Japan's military build-up on their mandated islands increased, so did their fear of being discovered. By 1937, with significant secret fortifications on Truk, Jaluit, and Saipan, among others, Japan was hugely skittish about anyone who appeared to be penetrating the waters or airspace around their mandated islands in the Central Pacific.

The United States and her allies, particularly the British, who had their own interests in the Central Pacific, had gotten wind of Japanese military activities in the mandated islands. As a result, they were desperate to find out the extent of Japanese military fortification in this area. With the specter of a major war in the near future, intelligence was badly needed.

In 1935, Pan American Airways initiated Trans Pacific flights using seaplanes, then dubbed, for various purposes, "flying boats." This immediately nonplused the Japanese, who did not like the idea of regular flights near or possibly even over their mandated islands. They were particularly concerned because they felt that the routes that Pan American was blazing could be just as easily used by military planes. By 1937, they were apparently beside themselves regarding the now regular and growing flights across the Pacific. There are indications that the Japanese embarked on a campaign of sabotage, disruption and outright attack against both Pan American Airways and the U.S. Government. (See chapter entitled "Dark Agenda.")

The Central Pacific was becoming a hell of a dangerous place.

Earhart, due to her great celebrity and the high esteem in which she was held everywhere, could be expected to go from country to country without difficulty, or arousing suspicion. She could take the occasional clandestine aerial photo and then return to the U.S. without anyone being the wiser.

After discussing several alternatives with Earhart, including mid-air refueling by U.S. military planes, Roosevelt and Earhart agreed that an airstrip on Howland Island would be the most practical answer to the aviatrix' need for a mid-Pacific refueling site between Lae, New Guinea and the Hawaiian Islands.

Additionally, Earhart's ostensibly civilian flight provided an excellent non-military pretext for the U.S. to build an airstrip on one of their own Pacific islands, in an area where militarization by any nation was a touchy matter diplomatically.

The United States had previously quietly annexed Howland Island, along with Jarvis and Baker Islands in early 1936. The government recognized the strategic importance of the three islands in a possible war and decided to take action.

On the recommendation of Secretary of State Cordell Hull, FDR issued a memorandum dated February 19, 1936, instructing State to prepare an order to place the three islands under the trust of the Interior Department. Up until then, the British had claimed ownership of the islands. However the Secretary of State felt, and the White House agreed, that if they just quietly annexed the islands, this might not be noticed by the Brits, who had problems of their own just then. This ploy worked.

The judgment of the government as to the strategic value of the islands, especially Howland, was confirmed at the beginning of WWII, when the Japanese, on their way back from attacking Pearl Harbor, viciously attacked Howland Island. The airstrip on the tiny island was destroyed and two of the colonists were killed. The island was rendered completely useless and the damage was never repaired.

On January 12, 1937, a Coast Guard ship left Honolulu, bound for Howland Island, and carrying the equipment and personnel to build an airstrip. The airstrip was completed barely in time for Earhart's planned takeoff from Luke Field in Hawaii.

LEGERDEMAIN

There are also indications that at least two other islands were provisioned as alternate landing sites, a standard procedure in such flights.[1] These were Hull Island and Canton Island, and as the reader will see later in this book, James Donahue found in his investigation that the British had an intelligence presence on Hull Island.

Earhart's flight afforded the perfect pretext for a covert, global, pre-war mapping and reconnaissance effort. And the plan nearly worked. The noted and beloved aviatrix transited smoothly from country to country, rarely being delayed by red tape. She and her navigator, Fred Noonan, usually were expedited through customs wherever they went. And more importantly, Earhart was able to easily take her assigned covert aerial photos throughout most of her flight, especially across the vast and little patrolled wastelands of Central Africa. Until she finally arrived in Lae, New Guinea, the only significant problem Earhart had reportedly encountered was a case of dysentery contracted somewhere in Central Africa.

But in the Central Pacific, the tables would turn. The U.S. Government had seriously underestimated the paranoia and ruthlessness of the Japanese. Given FDR's known fondness for human intelligence, this seems strange, as it is known that the U.S. had some information sources in the area.

One of them was the reports of travelers to or through the Japanese Mandate Islands. Another was a quasi-formal intelligence network in the area in the form of Guamanians, who were allowed by the Japanese to work on their installations on Saipan. When they returned to Guam, they would be debriefed by U.S. officials on their experiences.

As a result, one wonders how the government could have so misjudged the Japanese intentions after years of feedback from their various sources. There had also been a number of incidents over the years since the early 1920s, which had demonstrated the extreme danger of impinging on Japanese airspace or waters.

Marine Lieutenant Colonel Earl Ellis visited the Caroline Islands in an attempt to gain intelligence. He traveled briefly in the islands and then died there under questionable circumstances on May 12, 1923. Another officer sent to investigate Ellis' death, ended up following the trail to Tokyo, a trail which reportedly led to the infamous Black Dragon Society. Unfortunately, before he could return to make a more complete report, the officer perished in the great earthquake of 1923. A strange affair...

---

[1] Donahue, James A., "The Earhart Disappearanace: The British Connection" (Terra Haute, IN: Sunshine House, 1987)

David K. Bowman

In 1936, Willard Price, a member of the National Geographic Society staff, managed to gain official permission from the Japanese government to visit the Japanese Mandated Islands. Amazingly, he was able to stay in the islands for four months. While there, he made many observations which he later reported to the U.S. government, and which provided much information that was useful during WWII. He also encountered intense suspicion and secretiveness on the part of the Japanese, and was only just able to complete his visit in the islands without being arrested by the Japanese. They regularly read his papers and journals and followed him and his wife around. He later wrote a series of famous books, including one on his experiences in the mandated islands entitled "Japan's Islands of Mystery".

Price's observations on Truk, in particular, were controversial, as he insisted that he saw no military installations. However, the U.S. Government's information from its other intelligence sources contradicted Mr. Price's assertions, indicating a military buildup. Also, it is known that the Japanese did their best to conceal their military facilities and made their military personnel dress in civilian clothes. Therefore, it is quite possible that the Japanese were able to successfully fool Mr. Price about their military installations.

Later in the decade, two U.S. naval officers went into the same area on an intelligence mission and were captured by the Japanese. They weren't as lucky as Price and were executed. U.S. authorities later received quiet notification of the executions through diplomatic channels, but given the circumstances, weren't in a position to protest much.

Just before Earhart's flight through the Central Pacific, Japan had begun preparing for an invasion of mainland China, later known as the Second Sino-Japanese War after their invasion on July 7, 1937. Almost all of the Imperial Japanese Navy (IJN) forces were pulled out of the area to support the invasion of China. Unfortunately for Earhart, however, the Japanese had assigned some forces to the area, a move by the Japanese that was far from coincidence.

Ever paranoid about unfriendly eyes seeing their secret installations, the Japanese watched the entire Earhart world flight carefully in the media, and lay in vigilant wait, lest she violate their airspace.

When the hapless aviatrix took off from Lae, New Guinea on July 2, 1937, ironically toward the rising sun, she was flying towards the jaws of a waiting dragon.

Chapter 4

Busy 1960, Fred Goerner And Admiral Nimitz

1960 was an extremely interesting, if unusual time for investigations into the Earhart disappearance. Prior to that year, U.S. Air Force Officer Paul Briand conducted the only significant private investigation. But by early 1960, several more popped up. Just exactly why so many investigations into the Earhart disappearance occurred or started during that one year, isn't at all clear. Perhaps, in the scheme of things, the time had come.

Paul Briand published the results of his investigation, in that year, in his book "Daughter of the Sky". Mr. Briand had worked with Major Joe Gervais, USAF, during his investigation, and Gervais, with the assistance of Captain Robert Dinger, USAF had started another investigation, "Operation Earhart". A few years later, Captain Dinger, finding the investigation too time-consuming for him, bowed out of Operation Earhart, and a friend of his, air force officer Joe Klaas, took his place.

Moreover, Thomas Devine and his associates were conducting an investigation of their own. This was the result of Devine's experiences on Saipan during 1944, which will be detailed later in this book. (See "One Night on Saipan") In the summer of 1961, Devine found himself being tailed on a short trip to and from the home of Muriel Earhart Morrisey.

1960 was one hell of a busy year for Earhart investigations.

One of the most widely publicized investigations into the fate of Amelia Earhart, of 1960 or any year, was the one conducted by Fred Goerner, a CBS news radio reporter working out of San Francisco, CA. In 1960, Goerner launched his investigation after the account of a San Mateo resident appeared in the San Mateo Times. The woman's name was

Josephine Akiyama, and she had lived on Saipan in 1937 at the time of the Earhart disappearance.

One day in the summer of 1937, according to Mrs. Akiyama, she was riding her bicycle down the beach road near Tanapag Harbor on Saipan. She was bringing lunch to her brother-in-law, Jose Matsumoto. Matsumoto worked at the secret seaplane base the Japanese had built and were operating on Saipan. Just as Mrs. Akiyama reached the main gate of the facility, a large, twin engine airplane flew past overhead, disappearing in the area of the harbor. A short time later, when she reached the beach area, she found a large group of people standing around two Caucasians. Initially, she thought they were both men, but then someone told her that one was a woman.

"They were both thin and looked very tired," Mrs. Akiyama said. "The woman had short-cut hair like a man, and she was dressed like a man. The man, I think I remember had his head hurt some way".

Mrs. Akiyama was certain it was 1937 because it was the year she graduated from Japanese school. "I was eleven years old that year," she observed.

The guards then took the two people away, and not long afterward a rumor went around the island that the two had been executed.

That was Mrs. Akiyama's basic story, and it really grabbed Fred Goerner's attention.

Interestingly, at one point applications were pending by both the Goerner expedition and the Gervais expedition for permission from the U.S. Government to visit Saipan. For some reason the Goerner expedition was favored with permission to visit Saipan, while the Gervais expedition's application was denied. Thomas Devine, too, had applications to visit Saipan turned down in favor of Fred Goerner.

The San Francisco radio newsman's expeditions to Saipan pursued several strategies, the first being to locate Earhart's gravesite and then attempt to excavate it. Later, Thomas Devine would indicate in his books that Goerner had selected the wrong graveyard. But nevertheless, Goerner aggressively investigated his site, once he had found it, hiring a local bulldozer operator. The approach would have given any archaeologist hysterics, and at the end of multiple visits to the island, Goerner had bulldozed a fair chunk of real estate, without success.

Goerner's most successful strategy was to locate and question eyewitnesses who had lived on Saipan in 1937.

Goerner developed much interesting new information regarding Earhart's fate, but failed to locate the Lockheed Electra. Although he brought up a vintage aircraft generator from Tanapag Harbor, later examination by Paul Mantz and representatives of the Bendix Corporation showed that the generator had come from a Japanese "Betty" bomber and not Amelia Earhart's Lockheed Electra.

Near the end of Goerner's investigation, he went to Washington DC to interview some government officials. The night before he left, he received a phone call from Admiral Chester Nimitz, a 5-star fleet admiral, one of the best the U.S. Navy ever had. Goerner had been in recent contact with the admiral during his investigation. What the Admiral told the newsman left him stunned but gratified: "Now that you're going to Washington, Fred, I want to tell you Earhart and her navigator did go down in the Marshalls and were picked up by the Japanese".[2]

This single revelation alone should have laid to rest a lot of questions, but, hardly any writer has even acknowledged this assertion in the intervening years. Although there are facts that seem to contradict the admiral's statement to Fred Goerner, the statement appears to be valid and merits more consideration than it has received.

Shortly after Goerner's book was published, yet another expedition set out for Saipan in search of the grave of Amelia Earhart. This expedition was led by Don Kothera and Ken Matonis.[3] Just after WW II in 1946, Kothera had been stationed on Saipan. During his off-time, Kothera and a friend had occasion to stumble across the wreck of a civilian type plane near a gun emplacement in the jungle.

Kothera remarked on the shiny exterior of the fuselage: "...in mint condition, just like a silver spoon."[4] He noted also that the wings and engines had been "mechanically" removed, unlike those of the other planes which had been sheared or blown off. Kothera had his picture taken in front of the civilian aircraft fuselage that he had noticed among a huge number of wrecked planes. At the time, Kothera thought little of the plane, but years later, after seeing a photo of Earhart's Lockheed Electra, he thought there was a marked similarity with the plane he had found in the Saipan jungle.

Finally, encouraged by friends and family, Kothera set out for Saipan with several friends to find the airplane he had seen in 1946. Remembering the photo once taken of him in front of the aircraft fuselage, searched his possessions thoroughly, but to his frustration, could not find the photo.

---

[2] Goerner, Fred, "The Search for Amelia Earhart", (NY: Doubleday, 1966, pg. 305)

[3] Davidson, Joe, "Amelia Earhart Returns to Saipan," (Bloomington, IN: Unlimited Publishing, 3rd Ed. 2002)

[4] Howard, Michael, "Raiders of the Lost Aviatrix," Plane Dealer Magazine, Cleveland Plane Dealer, December 5, 1982

He would not have this crucial piece of evidence to help confirm that the plane he had found was Amelia Earhart's Lockheed Electra.

It turned out to be a much harder task than they thought. The expedition failed to find the airplane wreck, but ended up identifying a gravesite through talking with local eyewitnesses. Eventually, they succeeded in excavating the gravesite, recovering a number of human remains. Later examination by a forensic specialist determined that the remains were of both a male and female. The female was calculated to have been approximately 40 years of age. Moreover, analysis of a gold bridge found with the remains, found that it dated to the 1930s. While suggestive, the findings were not conclusive. Nevertheless, Kothera, Davidson, et al, continued to believe that the remains were those of Amelia Earhart and Fred Noonan. Were they?

Later in their investigation, the Kotheras had the opportunity to talk with Margot DeCarie, Amelia Earhart's secretary during the last few years before her disappearance. DeCarie reminisced that in March 1937 after the Luke Field crash, she had been sent by Earhart to the Burbank Airport to pick up high level presidential advisor Bernard Baruch and Eugene Vidal, as well as an assortment of admirals and generals. When the Kotheras asked DeCarie what was discussed, she remarked wryly, "I know one thing, they didn't' come out here for her autograph.

But DeCarie had some more provocative to say. She that just before Earhart took off for Miami, she gave DeCarie a large, well-filled manila envelope, telling her secretary, "If I don't come back, destroy it without opening it."

DeCarie stated that she felt she had been given the task because Earhart had thought that her own mother or GP would have been overwhelmingly tempted to look at the material. After the aviatrix' disappearance, DeCarie went to her parents' home and burned the envelope in front of them in the fireplace.

"What makes me boil," DeCarie concluded, "is that she didn't get any recognition from her government. You know, they always say a person who does intelligence work for their government might as well sacrifice their life and they're not going to get any praise for it."

DeCarie's remarks were tremendously tantalizing. They made it clear that she believed Earhart had been a on secret mission for the government and had been "hung out to dry" for political reasons. Her most intriguing remarks were about the mysterious envelope she had been given to hold by her employer and later destroyed.

What was in that envelope that it couldn't be disclosed to anyone—not even Earhart's private secretary?

A telling event occurred during a visit which Florence Kothera made to Washington DC in 1969 as part of the Kothera investigation. During that visit, Mrs. Kothera went to the Japanese Embassy. The ambassador was not in that day, an aide agreed to see Mrs. Kothera, who promptly and candidly told him of the Kotheras' investigation and asked him if he could confirm that Amelia Earhart had been beheaded by the Japanese on Saipan in 1937.

The question greatly consterned the aide, visibly disturbed, who stiffened noticeably in his chair. Mrs. Kothera then asked the aide what difference it could possibly make to confirm that event at that late date.

"Of course it makes a difference" the aide roared, getting to his feet and placing his hands on his desk. His face darkened with emotion, he continued, "It does make a difference, she was a citizen of the United States!" Abruptly he said down, his gaze averted. Mrs. Kothera finally had no choice but to leave at this evident dismissal. Thirty-two years after the disappearance of Amelia Earhart, the incident was an explosive indicator of the huge sensitivity of the Japanese over the subject. They clearly knew something they did not wish to discuss publicly under any circumstances.

Chapter 5

Crash Near Bicycle Lake

On the evening of December 15, 1961, a group of civilian and military pilots came together for the annual Lockheed pilots' party at the Palmdale Country Club in California. As the evening wore on, many of the pilots began to swap stories of daring-do and soon challenges to perform even more daring deeds began to pass between them. At approximately midnight, a group of eight pilots left for nearby Quartz Hill Airport, where two of their group, Charles Kitchens and Braxton Harrell, had decided to take off on an impromptu flight to Las Vegas. The aircraft in question was a Lockheed Electra.

Kitchens was the head of a three-man consortium that had purchased the Lockheed ship on February 14th of that year from Paul Mantz, Amelia Earhart's technical advisor. The craft was later described by a mechanic who worked at the hangar as being the "...most beautiful Electra I had ever seen. Not a scratch or dent on it. It appeared to have not more than 500 hours total [flying] time. It had seen a lot of storage in its day".

At 3:00 a.m. on December 16, 1961, the Lockheed Electra flew over the Bicycle Lake Army Airbase and made a turn to the left, evidently getting into pattern to land on Runway 22, which was lighted and available.

A moment later, at 3:01 a.m., the beautiful, mint condition craft was observed by base personnel to hit the side of nearby Mt. Tierfort, explode and burst into flames. The ensuing official investigation revealed that the port (left) engine had failed, causing a serious power loss and making it impossible for the crew of the ship to avoid hitting the mountainside. The report also indicated that this aircraft did not have featherable propellers (an important feature in making a safe emergency landing). Coincidentally, neither did Amelia Earhart's Electra, as full-feathering, constant speed propellers were not introduced on Lockheed aircraft until September 1937, shortly after Earhart's disappearance.

Subsequently, Major Joseph Gervais, who had been conducting his own investigation into the disappearance of Amelia Earhart, "Operation Earhart," got wind of the crash. Securing the permission of the base commander of Fort Irwin, the reservation on which Bicycle Lake AAB was located, Gervais climbed Mt. Tierfort to inspect the site of the crash. An experienced aircraft crash investigator, Major Gervais was well qualified for this task.

At the crash scene, Gervais found that some pieces of the Electra still remained. Strangely, identifying plates of all sorts had been carefully removed from the wreckage. However, part of the CAA registration number survived on an undamaged portion of the fuselage. Gervais took photos of this. Then, Gervais began searching further from the main crash site, having been taught by his crash investigation experience that other pieces of an aircraft are usually thrown some distance from the impact site.

A hundred yards from the main impact site, in a rocky ravine, Gervais found three sections of exhaust manifold, which represented parts of both airplane engines. Using a hammer and chisel, he broke open the cowlings and discovered that the identification plates for the manifolds were intact.

Each plate was stamped: "Exhaust Manifold manufactured by Solar Aircraft Corp., Lindbergh Field, San Diego, California. Model 12. Delivery Date May 13, 1937". Gervais determined that the crashed ship was a Lockheed Electra Model 12A, Serial No. 1243, FAA Registration No. N16020, the same number assigned to the Lockheed aircraft flown by Amelia Earhart during her disappearance.

Chapter 6

Operation Earhart

"Find Amelia Earhart's airplane, and you'll learn what happened to Amelia," Joe Gervais, a retired U.S. Air Force officer and Earhart researcher declared at the beginning of his 1970 book "Amelia Earhart Lives".

This is one of the most astute remarks that has ever been made about the search for aviatrix Amelia Earhart in the 67 years since she went missing on a stormy morning in the central Pacific on July 2, 1937. It became a central theme of Operation Earhart in the mid 1960s, propelled by the crash at Bicycle Lake.

At the beginning of "Amelia Earhart lives," Klaas and Gervais disclosed the crash of a Lockheed Electra in California on December 16, 1961 in the wee hours of the morning. What was so significant about this crash was that the aircraft bore the federal aircraft registration number of Amelia Earhart, N16020. According to the history books, this crash could not have happened—this craft was supposed to have disappeared over the Central Pacific in 1937.

If N16020 did not go down with Amelia Earhart in 1937, how did it end up on a hilltop in California in 1961?

When Gervais later consulted FAA records at the FAA Records Center in Oklahoma City, OK, he found a partial answer to the question, although he and Joe Klaas didn't recognize at the time the true explanation of what they had discovered.

Gervais was stunned to discover that there were TWO craft to which the number 16020 was issued: NR16020, Model 10E, Serial No. 1055, and N16020, Model 12A, Serial No. 1243. Since under FAA regulations, the same number is never assigned to more than one aircraft, Gervais was greatly puzzled.

Moreover, Gervais discovered that the second Electra, Serial No. 1243, was involved in a series of transactions between 1940 and 1961, culminating in the sale of the aircraft to an aircraft investment company in California shortly before its destruction in a crash at Bicycle Lake in December 1961.

Incorrectly, for some reason, Gervais confused the sequence of sales transactions of the Model 12A Electra. In his 1970 book "Amelia Earhart Lives", Gervais reported that the Model 12A Electra was sold by Canada in 1940 to the Charles Babb Co. and six years later to Paul Mantz after a series of peculiar, closely spaced transactions in 1946.

What I found after nearly a year of correspondence with the FAA was somewhat different. (See Appendix VI) In mid August 2004, I received a large package of document copies relating to the Model 12A Electra and found that the aircraft was manufactured in late 1937, with a license being granted on December 4, 1937 in the name of Lockheed Aircraft Company.

The following spring, the aircraft was sold to Western Air Express, who held the craft until June 1940 when they sold it to Charles Babb Co. in New York City. Charles Babb turned around and immediately sold the craft to the Canadian Department of Defence.

The Canadian government held the craft, apparently without its ever been flown, until after the war.

In January 1946, the Model 12A Electra changed hands again, in a series of three closely spaced, non-sequential transactions, ending up in the possession of Paul Mantz. CAA records showed that on January 21, 1946, Paul Mantz closed a deal to buy the Electra from Edward Ahr, of Timmins, Ontario. The aircraft was listed as war surplus. This seems to have been an advance deal, as Edward Ahr didn't buy the Electra until four days later, on January 25, 1946.

Meanwhile, on January 24, 1946, CAA records indicate, Algoma Air Transport bought the craft from the Canadian War Assets Corporation, for the oddly low price of $1.00. The next day, January 25, 1946, Algoma Air Transport sold the aircraft to Edward Ahr for the equally low price of $1.00, thus bouncing ownership back into Canada. This made it possible for Edward Ahr to consummate their deal with Paul Mantz and transfer ownership to him, for the far more substantial price of $20,000.

The above was intriguing as hell to me, but after reviewing all the document copies sent by the FAA, there was simply no connection between Amelia Earhart and the Lockheed Electra Serial No. 1243. During its complicated odyssey across two borders, the Model 12A Electra, with manufacturer's serial number 1243, which began its life as NC18955, went through a series of number changes. After it was purchased from the Canadian government, it was re-designated NC60775, and later was assigned Amelia Earhart's old registration number at the request of Paul Mantz.

It seems that Paul Mantz did indeed purchase the Model 12A Electra for use in a projected motion picture, which never materialized.

Disappointingly, the promising lead offered by the crash at Bicycle Lake has petered out. But Joe Gervais made his most important discovery when he inadvertently encountered Irene Bolam at a luncheon in the summer of 1965 (see "The Bolam Affair"), and made further headway when he uncovered the mystery of Wilbur Rothar (see "The Bizarre Tale of Wilbur Rothar").

Another possibility that Operation Earhart ventured was that Earhart's freedom from Japanese captivity was secured through secret negotiations with the emperor of Japan at the end of the war. It is an interesting historical fact that the emperor got to keep his throne after the war, while numerous subordinates were hung by war crimes tribunals. Gervais and Klaas asserted that the emperor released Earhart in return for his keeping his power.

It was and is a plausible scenario. The problem is that the only proof of it so far has been an ambivalently worded formerly confidential communication from the emperor, indicating that he would like to keep his power after the war. So it appears that he was maneuvering for a deal. But thus far, no other details have surfaced regarding those negotiations.

When "Amelia Earhart Lives" was published in 1970, it created an enormous stir. Subsequently a lawsuit was filed and, for a time, the book was not available to the public. It was an amazing phenomenon, in which Klaas and Gervais were denounced to the point of vilification.

In some ways it reminds this writer of the furor that occurred after New Orleans District Attorney Jim Garrison announced his prosecution of suspects in the Kennedy Assassination in the 1960's. Garrison was not just criticized, but was savagely denounced by the media and establishment in general. Subsequent developments have vindicated Garrison, showing that the defendants were definitely involved in the Kennedy Assassination and that high level members of the establishment were also involved in the conspiracy. Jim Garrison had proceeded correctly. As a result, whenever this writer notes savage denouncement of anyone, without any visible substance or foundation, he starts to wonder. Because for some reason, frequently, the truth seems to draw more fire than falsehoods.

Chapter 7

The Bolam Affair

During the final phase of Joe Gervais' investigation, Operation Earhart, events continued to take strange turns. In the summer of 1965, Gervais received a letter from Viola Gentry, a friend of his aunt, inviting him and his wife to a special luncheon in their honor given by the Early Flyers Club. The meeting was to be held at the West Hampton Air Force Base and Major Gervais was scheduled to make a presentation to the club members regarding Operation Earhart. Miss Gentry promised Gervais that many of Amelia Earhart's old friends would be there.

Gervais cheerfully accepted the invitation, little realizing the experience awaiting him.

On the day of the luncheon, Gervais and his wife met Viola Gentry in a large banquet room at the Sea Spray on the Dunes in East Hampton. Gervais had brought along a camera loaded with color film and was happily snapping photos of various old-time aviation figures like a sightseer as he strolled around the room.

Just then, Viola Gentry, who had been looking toward the nearby sea, turned back to look into the room and stopped to stare. Across the room, a silver-haired man and woman had entered through a door.

"Why, there's Irene Bolam," Viola said in awe. "It really is Mrs. Bolam".

Gervais glanced in the same direction and froze. The woman Viola Gentry had called Irene Bolam had a stunning resemblance to Amelia Earhart. After spending years of examining photos and film clips of Earhart, Gervais was almost thunderstruck. In fact, he immediately wondered if he was looking at a 68 year old Amelia Earhart.

Keeping his voice almost to a whisper, Gervais asked Miss Gentry with a stammer, "Viola, could I please meet that woman?"

"Oh yes," Miss Gentry replied. "You must meet Mrs. Bolam".

Miss Gentry led Gervais over to Mr. and Mrs. Bolam and introduced them to the air force officer. Gervais couldn't get over the deja vu-like feeling that he knew or had known Mrs. Bolam.

"I'm most delighted to meet you, Mrs. Bolam," Gervais said. "Were you a friend of Amelia Earhart?"

"Yes," Mrs. Bolam said quietly with a distant smile. "I knew her".

"I'll bet you knew Amelia rather well".

"Yes, I knew her rather well," Mrs. Bolam said, an enigmatic twinkle in her eyes. Pinned to her dress were several medals. Joe Gervais indicated in his book that they were a miniature Distinguished Flying Cross, (which can only be worn by those who have been awarded it), a miniature of the awarded to Amelia Earhart by New York for her transatlantic flight, and a miniature major's oak leaf cluster. Unfortunately it is hard to tell from the photo he took if this was correct.

"Were you a pilot, Mrs. Bolam?" Gervais asked.

"Oh, yes," she replied softly.

"Did you ever fly with Amelia Earhart?", Gervais continued.

"Yes, Major. I flew with Amelia," Mrs. Bolam said.

At this point, Gervais began talking with Mr. Bolam, suddenly afraid of appearing too obvious.

Near the end of the conversation, Gervais asked Mrs. Bolam for her address so that he might write to her some time. After exchanging looks with her husband, Mrs. Bolam gave Major Gervais a card with her address on it.

Gervais then stepped back from the couple and asked if he could take a picture. Mrs. Bolam started to protest and then Gervais snapped the shutter.

"Just one," Gervais said. "I'll send you a copy".

Irene Bolam stared in confusion at her husband. Viola Gentry gasped and Guy Bolam shrugged.

"Oh, well, I suppose just one," Mr. Bolam conceded.

"Are you a Ninety-Nine?" Gervais then asked Mrs. Bolam, to which she replied "Yes". He also asked her if she was a Zonta, a feminists' sorority to which Amelia Earhart had belonged and Mrs. Bolam again replied in the affirmative.

At that point, Guy Bolam abruptly made an excuse and he and Mrs. Bolam moved to another part of the room.

After lunch, Gervais gave a talk about Project Earhart, at the conclusion of which there was a brief surprise ceremony in which his wife was presented with the Amelia Earhart Award for Outstanding Contribution to Research in the History of Aviation, for her assistance to her husband in Operation Earhart. Gervais noted that Mr. and Mrs. Bolam had left before his talk.

That evening after Gervais had returned to his hotel room, he received a call from Irene Bolam, inviting him and his wife to her home the following evening. When Gervais asked the nature of the invitation, Mrs. Bolam told him she wished to discuss his investigation into the disappearance of Amelia Earhart.

Gervais had to decline the invitation because he had to pick up his children the next day from a relative's home and had airline seats reserved for the next day.

He would later realize that declining Irene Bolam's invitation was one of the biggest mistakes he ever made, as he spent the rest of Operation Earhart trying to get in contact with Mrs. Bolam and she failed to keep each of the appointments which she subsequently made with Gervais.

After the Long Island luncheon, Gervais and Klaas made repeated attempts to check Irene Bolam's background finally writing a letter to Mrs. Bolam asking her to provide proof of her identity so that they could follow other leads if they were wrong.

Mrs. Bolam wrote them back, giving Gervais' friend, Viola Gentry, as a reference, as well as one Elmo Pickerill of New York State. Viola Gentry send a fairly brief note back to Gervais, but Mr. Pickerill sent a more extensive letter detailing Mrs. Bolam's background.

Unfortunately, Klaas and Gervais could find no independent source with which to verify Mr. Pickerill's letter. A query to the FAA Records Center in Oklahoma City, OK, brought the response that Ms. Bolam had been issued a student's license (no number), on 9-20-32 and a private license, #28958, on 5-27-33.

Subsequently, Gervais and Klaas requested a copy of Mrs. Bolam's pilot's license and were sent a copy of a non-commercial pilot's license, #28958, dated 5-31-37. The date had been crossed out and penciled in above it was 6-1-37. The address given for Mrs. Bolam was in Brooklyn, NY and the license was unsigned.

Clearly, the copy of the license contradicted the information the FAA had given Gervais in their first letter. Why would such a contradiction exist? Was this evidence of a cover-up?

One new bit of information that has surfaced among researchers is that in the spring of 1937, Irene Craigmile's 1933 license was about to expire and the 5-31-37 license may have been automatically issued, as a matter or course. Or issued in error.

If the license was automatically issued, in contemplation of Ms. Craigmile showing up to claim it, the timing is remarkably coincidental. For the document was dated the day AE left on her world flight.

It is a confusing matter that is far from clear.

Gervais and Klaas wrote another letter to Mrs. Bolam, this time asking her about not only her past, but the purpose of Earhart's last flight, and Wilbur Rothar as well (See chapter on Wilbur Rothar). They never received an answer to that letter.

Much later, in one of Gervais' last telephone contacts with Irene Bolam, she made some telling remarks. After Gervais implored her to arrange an appointment for them to talk, Mrs. Bolam said, "Oh, I can't see you in this country".

"I beg your pardon?" Gervais said.

"I mean, I couldn't meet with you in the United States. Look, Major Gervais. I once had a public life. I once had a career in flying. But I've retired. I've given that all up now. As a major retired from the air force, you should be able to understand this".

She proceeded to make an appointment to meet Gervais at a hotel in Montreal, which she didn't keep, to Gervais' frustration.

Subsequently, after Gervais had returned to his home in Las Vegas, Viola Gentry flew there from the East Coast. The next day, Gervais had dinner with Miss Gentry and two other persons in a Las Vegas restaurant.

"Irene has gone to Paris. You'll never see her again," Miss Gentry told Gervais. Ironically, Gervais and Klaas would indeed see Irene Bolam again one day, in open court.

"You know, Viola," Gervais said at length "there are a lot of people interested in this case. It could be worth a lot of money to find out what happened out there on July 2, 1937".

Nodding, Miss Gentry said "That's what Amelia says".

Everyone at the table stopped talking and stared at Viola.

"Viola, do you realize what you just said?" Gervais asked softly.

"What?" Miss Gentry said.

"You said that's what Amelia says. As if she were alive".

"Oh, my. Did I say that?" Miss Gentry said. "I meant Muriel. You know. . . Amelia's sister. Muriel Morrissey. I often confuse their names".

After that, Joe Gervais was never able to speak with Irene Bolam again. And, interestingly, when he checked with the Ninety-Nines and the Zontas, he was told that the name "Irene Bolam" did not appear on their membership lists.

LEGERDEMAIN

Thinking about it recently, this writer had one thought: The membership lists of the two old-time aviation clubs may not have included the name of Irene Bolam, but there was one way Mrs. Bolam could have been a member of those two clubs. Their lists did include the name of Amelia Earhart . . . One strange postscript to the Bolam affair was the result of the lawsuit she later filed against Mr. Klaas and Mr. Gervais when "Amelia Earhart Lives" was published. When the suit reached court in 1975, the court asked Mrs. Bolam to give her fingerprints to the court in front of the judge to confirm her identity and settle the matter. This Mrs. Bolam declined to do. According recent information from Joe Klaas, the court then recommended to Irene Bolam's attorney that perhaps they ought to negotiate a settlement over the case with McGraw-Hill's attorneys.

Subsequently, this is what happened. A confidential settlement with worked out with McGraw-Hill and Klaas and Gervais had no further involvement in the case. They were even indemnified, according to Joe Klaas, from further lawsuits by Irene Bolam. Shortly afterward, she moved to Paris, France.

Yet another strange postscript to the Bolam affair was an investigation conducted by screenwriter Tod Swindell in 2002[5]. (See chapter on Col. Reineck.) Swindell arranged for two forensic pathologists to use the recent technology of superimposition of photos.

In this case, photos of Amelia Earhart and Mrs. Bolam were superimposed over each other and then closely studied to determine whether the bone structure coincided or not. After comparing the bone structure in Mrs. Bolam's face and hands with those of Amelia Earhart, the pathologists concluded that the bone structure coincided exactly.

John Bolam, half brother of Mrs. Bolam's husband, Guy Bolam, stated at first that "we were inclined to think Irene probably was not Amelia Earhart". "However," he went on, "the forensic studies are very convincing". Bolam also stated that while Mrs. Bolam denied being Amelia Earhart, she was not an ordinary housewife as she claimed. "She was influential, knew many well-placed people and was well-traveled".[6]

In the end, one is left to wonder, in light of the above, why Mrs. Bolam, wanting to keep her real identity, whatever it was, confidential, would have subjected herself to the publicity of a lawsuit. It is a mystery that only time may solve.

---

[5] "65 years later, the mystery of Amelia Earhart continues," Ron Staten, AP.
[6] Ibid.

Chapter 8

The Enigma of William Van Dusen

During Joe Gervais' pursuit of Operation Earhart, he had a series of contacts with a gentleman who was a vice president of Eastern Airlines, Mr. William Van Dusen. The odd thing is that Major Gervais did not initiate the series of contacts with Mr. Van Dusen. It was Mr. Van Dusen who started the series of fascinating, if enigmatic contacts.

One day in May 1964, Gervais received a large envelope from Eastern Airlines, which contained a series of documents with a cover letter signed by William Van Dusen's private secretary. The text of the short letter from the secretary was as follows:

"Dear Major Gervais:
"On March 12, I wrote to Colonel Robert G. Wilson at the Office of Information in Washington, DC regarding the search for information about Amelia Earhart and was informed by him that you would be the one to help.
"I am attaching copies of all correspondence so that you will be able to follow exactly what has happened.
"Your cooperation and early reply would be greatly appreciated.
Sincerely,
Shirley J. Hancock (Miss)
Secretary to
William Van Dusen"

The attached documents were copies of letters that detailed the results of an inquiry made by Mr. Van Dusen into the disappearance of Amelia Earhart. Included were copies of letters Mr. Van Dusen wrote in reply. Mr.

Van Dusen, quite clearly, wished to know what Joe Gervais had uncovered in his investigation.

Already having experienced problems due to having been too candid at the wrong time, Gervais did not respond to the letter, but instead, made extensive inquiries about the backgrounds of William Van Dusen and Fred Noonan. Interestingly, when he wrote to each man's home town, he found that no birth certificate was on file for either. It was not possible to find hard proof of where or when each man was born. Gervais began to wonder if they could be one and the same individual. As Operation Earhart went on, Gervais would have other experiences that would make him wonder even more.

In early 1965, Gervais received an invitation to a luncheon in honor of him and his wife (see chapter entitled "The Bolam Affair") to be held on Long Island on August 8, 1965. Arriving in New York a day early, Gervais made a brief visit to William Van Dusen's estate in Connecticut.

Upon meeting Van Dusen, Gervais was immediately struck by the startling resemblance between Van Dusen and Fred Noonan. Mr. Van Dusen showed a strong interest in Major Gervais' investigation, telling Gervais that he had been a friend of Amelia Earhart.

"What makes you think Amelia Earhart didn't just miss Howland Island and crash in the ocean?" Mr. Van Dusen asked.

"With an experienced navigator like Noonan aboard?" Gervais said. "You knew him, didn't you?"

"I knew him," Van Dusen said with disdain. "Noonan was a bum".

"But a good navigator, nevertheless," Gervais countered.

"Noonan couldn't navigate his way across my duck pond," Van Dusen replied. "What other reasons do you have for not thinking they didn't go down in the drink?"

Gervais was startled at the air of authority in Van Dusen's manner, as though Van Dusen was speaking from a superior and knowledgeable position. Nonetheless, Gervais proceeded to brief Van Dusen on his efforts to date. When Gervais got to his investigation into Wilbur Rothar, Van Dusen offered to place a story regarding Rothar in two different New York newspaper columns to see if any leads could be developed. Subsequently, Gervais found that Van Dusen had kept his promise, when the stories actually appeared in the paper. Disappointingly, Van Dusen later told Gervais there were no responses to the stories.

All during the visit with Mr. Van Dusen, Gervais occasionally asked questions regarding Van Dusen's involvement with Amelia Earhart. Each time, the older man adroitly changed the subject without really answering the question.

Just before dinner, Mr. Van Dusen's attractive young wife abruptly said, "Why don't you tell Major Gervais what he wants to know?"

Van Dusen darkened over for a moment and barked at his wife, "Let's see you work your way out of this one". After a tense moment, Van Dusen subsided and was once again the suave host.

During dinner Gervais brought up the subject of Fred Goerner's investigation and Van Dusen laughed.

"Goerner isn't even looking for her in the right ocean".

That was a remarkable statement. What did Van Dusen mean by it?

Again Gervais was startled at the ambiance of absolute knowledge in Van Dusen's voice. There was some further brief and enigmatic small talk and shortly afterward, Gervais took his leave. At the doorway as he was leaving, Gervais abruptly asked, "Bill, do you ever have any trouble proving who you are?" Van Dusen's face didn't change expression an iota, as he said, "What do you mean?"

Casually, Gervais told Van Dusen, "You ran a search on me all over Washington, so I ran my own on you. But I can't find a birth certificate on you in Toledo, Ohio. That where you were born?"

Smiling, Van Dusen replied, "Yes, as a matter of fact . . . sometimes I do have a little trouble proving who I am. Be sure to check back with us after your talk tomorrow".

As Gervais prepared to drive away, he looked over toward the doorway of the house. Van Dusen was standing there next to his young wife, looking just life Fred Noonan did in all those old photos, a deep crease in his cheek and his hands on his hips.

Later, after the luncheon at which he met Irene Bolam, Gervais dropped by Van Dusen's office at Eastern Airlines. Van Dusen had no news, but promised to advise Gervais if he received any responses to the newspaper stories he was going to have run.

Gervais handed Van Dusen Irene Bolam's card. "Do you know this lady?" he asked.

"No," Van Dusen replied.

"You ought to meet her. She's a very nice lady and was a friend of Amelia Earhart's," Gervais said.

"I'll look her up," Van Dusen promised and then changed the subject.

A short while later as he was leaving, Gervais asked Van Dusen, "Are you going to look up Mrs. Guy Bolam?"

"Yes," said Van Dusen.

"Well, don't you want her address and phone number then? They might not be listed in the phone directory". Gervais offered Van Dusen Irene Bolam's card.

LEGERDEMAIN

Van Dusen said, "Oh, yeah, let me copy that," taking the card. His tone and manner made it seem that he was going through the motions for Gervais.

Gervais wondered if Van Dusen already had Irene Bolam's address and phone number.

A short while later that same summer, Gervais put together a trip to the South Pacific to visit a place called Winslow Reef, a series of sand-covered reefs about 170 miles southeast of Howland Island. Gervais was convinced that the reef would yield some answers to the Earhart mystery.

Unexpectedly, the ship he had hired had run into serious engine trouble and taken on a lot of water. Gervais got the news in a series of telegrams and a letter from the ship's owner. Repairs were expensive enough and time consuming enough that the ship would not be available for that summer. Gervais was forced to postpone the trip to the following summer.

The following summer, in 1966, Gervais was busily trying to get together another trip to Winslow Reef, when he had a surprise visit at his Las Vegas home from William Van Dusen.

Van Dusen told Gervais that he had had dinner the previous evening in California with Jackie and Floyd Odlum. The Odlums were old-time aviation figures in their own right and had been close friends of Amelia Earhart. They were, according to Van Dusen, very interested in the progress of Gervais' investigation.

Van Dusen then proceeded to comment on the great danger involved in the Winslow Reef expedition, with the modest equipment and boat, which was all Gervais could afford to hire. He suggested that the Odlums were interested in financing the expedition and could provide Gervais a large, powerful boat equipped with a helicopter.

Gervais was quite enthusiastic.

Van Dusen asked Gervais if he would be willing to make an audio tape explaining what his investigation had found and what they hoped to find. He would then play it for the Odlums to persuade Floyd to finance the Winslow Reef trip.

"You think he would?" asked Gervais.

"If you can convince him on tape the expedition stands a good chance of settling this Earhart thing once and for all. You'll make the tape?"

Gervais promptly made a tape for Van Dusen and Van Dusen left. After Van Dusen's departure, Gervais then called his friend and research partner Lt. Colonel Dinger.

"Did you mention your contact with Mrs. Guy Bolam on that tape?" Dinger asked.

Gervais replied in the negative, telling his friend that he only wanted Van Dusen, et al, to think that Winslow Reef was their main concern. Dinger warned Gervais that Van Dusen and the Odlums might simply be trying to find out everything Operation Earhart had uncovered. Gervais agreed and said that was the reason for his approach on the tape.

Gervais heard nothing further from Van Dusen regarding the tape.

Many months later, near the end of Operation Earhart, just after Gervais had completed some research on Wilbur Rothar in New York, he took the opportunity, while he was in town, to visit William Van Dusen. Not long before, Gervais had a made an appointment to meet with Irene Bolam in Montreal (see chapter entitled "The Bolam Affair").

Early during the dinner that night, Mr. Van Dusen casually said to Gervais, "I hear you are going to Montreal, tomorrow".

Gervais was extremely surprised, as he had only recently made the appointment to meet Irene Bolam at a hotel in Montreal and as far as he knew, only he and Mrs. Bolam knew of the appointment.

"How did you know that?" Gervais asked in amazement

"Smoke?" asked Van Dusen, ignoring the question. He produced with a great flourish, a silver cigarette case. This case had clearly belonged to Fred Noonan, as evidenced by an inscription on the case from Noonan's Pan American Airlines colleagues. The case even contained an ancient cigarette, which Gervais recognized as being of 1930s vintage. When Gervais asked Van Dusen where he had gotten the case, the enigmatic airline executive merely said, "Oh, I don't know. I guess Fred must have left it here or something".

Gervais tried a couple of more times to get Van Dusen to discuss the cigarette case, but the Eastern Airlines executive just wouldn't say another word about it. That was the last contact that Major Gervais had with Mr. Van Dusen. Van Dusen's remark about the trip to Montreal was a big giveaway that his assertion that he did not know Irene Bolam was not true. He could only have found out about Gervais' appointment via contact with the mysterious matron. Mr. Van Dusen was very interested indeed in Major Gervais' movements.

Chapter 9

The Bizarre Tale of Wilbur Rothar

Following the disappearance of Amelia Earhart, a strange series of events occurred in August 1937, which remain one of the most remarkable features of the Earhart disappearance, and may also be an important clue as well.

On July 25, 1937, newspapers reported that Sydney S. Bowman, head of the Pan Pacific Press Bureau and close friend of Putnam, had posted a $2000 reward on Putnam's behalf the previous day. The reward was for information "which would definitely clear up the mystery surrounding the disappearance of Amelia Earhart and Fred Noonan".[7] Bowman was quoted as saying the reward would be paid for the "recovery and delivery of any part of the Earhart plane or its contents, which may be identified and which would clearly reveal the fate of the missing flyers".

This convoluted statement sounded as if it had been written by an attorney.

"There is a chance that further investigation may disclose information upon which absolute conclusions may be reached," Bowman went on. "It is regarded as possible that ships operating in the south Pacific may come upon evidence that would end speculation regarding the aviatrix and her navigator, and it is to encourage this cooperation that the reward has been offered".

On August 5, 1937, a story appeared in the New York Times: "Janitor is Seized for Earhart Hoax".

---

[7] "Reward Offered for Aid in Clearing Earhart Mystery," Coshocton Tribune, Coshocton, OH, 7-25-37, front page.

The article reported that several days before, on Saturday, George Putnam had received a strange note, which had been left for him at the Hotel Barclay. The Barclay was a place Putnam habitually stayed when in town. The note read: "We have your wife on the ship. I will call Sunday at 2 o'clock".

About 10:00 p.m. the same day, Putnam received a telephone call from a man, telling him that if he would arrange for a meeting the next day, Sunday, he would receive important information about Amelia Earhart. A meeting was promptly set for Sunday at 2:00 p.m. in Putnam's hotel room.

At 2:00 p.m. the next day, Wilbur Rothar reportedly showed up at the Hotel Barclay to tell Putnam an astonishing tale. Introducing himself as Mr. Johnson, Rothar then told the publisher that he had been employed on a vessel, which was running guns to Spain. The vessel was traveling from New Guinea to Panama. Several days out of New Guinea, the ship anchored off of a small island to take on a fresh water supply. In a cove on this island, according to Rothar, the landing party from the ship spotted a crashed airplane.

Nearby, was a woman who was delirious and very ill, staggering along the shore. According to Rothar, the woman was wearing only a pair of shorts. On the wing of the plane lay the body of a man who had been killed by sharks. Rothar said that the woman was taken aboard the ship and treated by the ship's Chinese doctor, during the remainder of their voyage.

When the ship reached Panama, according to Rothar, they recognized the woman from newspapers as Amelia Earhart. They flew into a panic because they were afraid their illegal activities would be discovered if they tried to put Earhart ashore for help.

The ship then continued to New York, Rothar said, and he was elected by fellow crew members to approach Putnam. Rothar stressed that many of the men aboard the ship were "cutthroats" and had talked repeatedly about dumping Earhart into the sea. He indicated that Ms. Earhart was so ill that she would have to be taken off the ship and to a hospital soon.

Putnam then requested proof of the story and Rothar said he had proof and would appear at Mr. Putnam's office the next day with it.

On Monday, Rothar appeared in Putnam's office with a scarf, which Putnam's secretary recognized as belonging to Amelia Earhart. Rothar stated that the crew had definitely decided that they would not release Earhart unless they were paid the $2000 reward.

Putnam said he would happily pay the reward and asked Rothar to come back the next day for the money. The next day, Rothar again appeared in Putnam's office and was given $1000 cash. While he was at a bank cashing a check for the other $1000, Rothar was arrested. Rothar reportedly confessed to trying to commit a fraud on Putnam. Authorities were reported to have discovered that Rothar was a janitor from the Bronx and the father of eight children.

Rothar later told authorities that he had gotten the scarf some three years earlier at a Long Island airport, when he had gone there to try to get a glimpse of Earhart. As Earhart got into her plane, the media reported, the scarf fell to the ground and Rothar retrieved it. Rothar reportedly said that he had kept the scarf as a memento.

Joe Gervais investigated this incident as part of Operation Earhart, and was very interested in the incident due to its strangeness.

A short time after Rothar's arrest he was arraigned and ordered sent to Bellevue Hospital in New York for ten days observation. Two months later, on October 13 1937, Rothar was committed to the Matteawan State Hospital for the Criminally Insane, pending trial on his extortion charge.

What Gervais discovered from painstaking checking of records was that Rothar remained institutionalized for the next 24 years—and was never tried for anything! This seemed a strange outcome for a run-of-the-mill extortion charge.

Interestingly, after various investigations into Earhart's disappearance began to get publicity in 1960, Rothar, now listed as Wilbur Rokar, was transferred from Matteawan State Hospital to Harlem Valley State Hospital on April, 19, 1960.

Two years later, Rokar, was then transferred to Central Islip State Hospital on March 23, 1962. In October 1962, he escaped from the facility and remained at large for a year. The following year, he was returned to that hospital and a short time later, on October 2, 1963, Rokar was officially discharged, never to be heard of again.

Muriel Morrissey, Amelia Earhart's sister, recounted the above incident very differently. She maintained that the amount of money that Rothar demanded was $5000 and that the scarf, which Rothar presented to GP, had been found by him in Hawaii, possibly in March 1937. She characterized Rothar as a "shamefaced and frightened young man," as opposed to the 42 year old father of eight, which the media reported Rothar to be.

Mrs. Morrissey recounted that GP did not prosecute Rothar, exclaiming to Rothar that his wife would not even let him discharge a member of his household staff for starting a kitchen fire. According to Mrs. Morrissey, GP gave the man $50 for the scarf and admonished him to mend his ways in the future.

The reality of what happened was that shortly after Rothar was arrested, GP Putnam was present in court to press charges against the hapless Rothar.

In her book, "Whistled Like a Bird," Sally Putnam Chapman, the granddaughter of GP Putnam recounted essentially the same scenario as Mrs. Morrissey.[8]

In a further development, Joe Gervais checked exhaustively and could find no evidence that Wilbur Rothar had ever existed. When he traced down the address for Mr. Rothar, which had been published in the newspaper, he found that the address was non-existent, and would have been somewhere at the bottom of the East River . . .

In a further search for information, this author discovered that in 1940 an article had been published in True Detective magazine regarding the Rothar affair, and had the good fortune to locate a copy of that publication.

The article reveals other details that at once both deepen the mystery and suggest a solution. The first thing I noticed was that Rothar's name had transformed from "Wilbur" to "Wilber". Also, in this article, Rothar claimed that he wasn't engaged in smuggling arms, but engaged in "dope-running" aboard a "fast freighter". In addition, further details were reported regarding Earhart's attire.

"She didn't have a thing on," Rothar told Putnam with some embarrassment, "except…"

"Except what?" Putnam prompted.

"Why, just some underwear," Rothar said hesitantly.

"What kind?" Putnam persisted.

"Well, it wasn't the kind women wear," Rothar said.

"What kind, men's?" Putnam asked. "Go on, describe the shorts."

"Why, there's nothing much to describe," Rothar said. "Just little, tight-fitting cotton things. There's a name for them". He paused for a moment". "—oh, I know. Jockey shorts.

Putnam was reportedly shocked because he had indeed purchased some jockey shorts for his wife just before she left on her last flight. "They're sort of cute and homely," Earhart had told her husband laughingly. "They ought to be comfortable to wear under slacks for flying".

The article stressed that GP felt that he was literally the only person in the world who knew that fact. It concluded that he had declared that he was unable to find an explanation for Rothar's knowledge of that information:

---

[8] Sally Putnam Chapman, "Whistled Like a Bird," (NY: Warner Books, 1997)

"He still insists that no one knew that intimate little secret—and he would welcome a solution".

Still another small surprise was in store when I got to the very end of the article and read:

"Judge Freschi put this cold-blooded schemer away from the world for the rest of his natural life. And he will never torture a grieving man again. He is behind locked doors today, still the baffling, uncompromising man who made George Putnam's life an agony for three long days.

"Amelia Earhart is dead. So the courts of Los Angeles ruled in January 1939.

"And who can blame George Putnam for wondering and puzzling about the unanswered riddles in the fantasy of Wilber Rothar, the sailor who never went to sea".

This passage was surprising because Rothar's legal status at that time was that of a defendant undergoing evaluation to see if he was competent to stand trial for extortion. Moreover, the crime of extortion did not command a life sentence. At the moment the True Detective article hit the newsstands, Rothar was languishing in a mental institution and not a jail.

Admittedly, some of the discrepancies in the True Detective article may be due to artistic license on the part of the writer. However, the new spelling of the name, along with the additional details of the story, only serve to increase the suspicion that the basic news story was completely bogus.

Yet another new piece of information in the article was a close-up photo of a nondescript looking man, whom True Detective said was Wilbur Rothar. (See reproduction in illustration section.) The photo shows a middle-aged man, and not the "young man" of the Putnam family/Muriel Morrissey versions of the incident.

At first blush, Wilbur Rothar seems to be a man who knew "too much" and was silenced by being hustled off to a mental institution. This is supported by the fact that a level of knowledge was attributed to Rothar which only GP Putnam, by his own admission, possessed.

If only Putnam knew those personal details, therefore, it was most likely Putnam who supplied most of the scenario, which was recounted in the media. Given that fact, the fact that no one named Rothar ever existed, as well as the complete inconsistency of all recountings of the incident, the whole affair almost certainly seems to have been an elaborate ruse orchestrated by GP.

But why?

Going further, since everything that was publicized may have been fabricated, even the one "reality anchor" in the incident, the scarf belonging to Earhart, which was mentioned in news releases, may have been a fabrication too.

Unfortunately, some all too real person was incarcerated in a mental institution.

For another thing, when Earhart investigations began to be publicized in early 1961, Rothar, now known as Rokar, was twice transferred to different facilities before the end of the next year. This doesn't even come close to coincidence. And what did Rothar know? Rothar's lengthy, unjustified and illegal incarceration attests to the fact that he knew something about the disappearance of Amelia Earhart. Something happened to cause GP to act immediately and cover it up with a false news release.

Which raises another question about the affair: why the phony news release? If the whole affair had happened in private, there would have been no need for a cover-up. Therefore, something must have happened, probably in GP's office, which wasn't private. In other words, some sort of scene must have occurred with other people present, and caused GP to swing into immediate cover-up mode.

Checking through newspapers for information on the Rothar case, I ran across an article that, while not directly relevant to the Earhart case, may offer a clue as to the nature of the Rothar affair.

A newspaper headline for Saturday, May 13, 1939 declared in huge bold type: "PUTNAM IS KIDNAPED [sic]". According to the article, earlier that same day, Putnam was found bound and gagged in an uncompleted house near Bakersfield. At the time, Putnam was in the process of publishing a book called "The Man Who Killed Hitler". The author of the book was anonymous. Putnam's captors had reportedly made threats before the kidnap, including the suggestion that it would be "healthier" if he ceased printing the book. "They were quite decent to me," Putnam was quoted as saying of his kidnappers, who demanded to know the name of the anonymous author. "They didn't even rough me at all," Putnam added.

There are two significant points about this news story. One is that it has all the earmarks of one of GP's publicity stunts. It just doesn't seem completely believable, except as a promotional effort for GP's book, "The Man Who Killed Hitler".

The other point is that the general style of this news story is exactly like that of the Rothar affair: melodramatic and flamboyant. In short, the 1939 kidnap report matches the flavor and style of the Rothar report. Both news stories seem to be false and seem to come from the same creative imagination. Did GP fabricate both?

As to who Rothar might really have been, that, too, is still a mystery, although Gervais and Klaas, in their book, flirted with the idea that Rothar was actually Fred Noonan. They speculated that he had gotten back somehow from the Central Pacific and perhaps had attempted to tell whatever he knew about the Earhart disappearance. The basis for this was Rothar's reported remarks and ramblings in his sealed court record.

Since the whole reported scenario looks strongly to be a fabrication, and also since someone was actually incarcerated in a mental institution, perhaps Klaas' and Gervais' proposition may not be so far-fetched after all.

Purely as speculation, could Noonan have actually escaped from the Central Pacific and make it to New York to throw a scene in GP's office? Did he threaten to expose something embarrassing regarding Earhart's disappearance? Did GP then hastily have Noonan arrested using his influence, and pass the ragged and unrecognizable Noonan off on the authorities as a madman? And was the luckless navigator then relegated to 24 years of miserable, anonymous confinement in a mental institution?

Or had Wilbur Rothar just been someone unfortunate enough to stumble across some information about Earhart so sensitive that Putnam fell all over himself to rush the hapless witness into the nearest rubber room?

In the 1930's, unlike today, it was extremely easy to have someone inappropriately committed to an institution, as one only needed the signature of a single physician on a document certifying the victim as insane.

Ultimately, the Rothar case, as incredible as it appears, cannot be completely dismissed out of hand for two reasons: 1) The whole scenario as reported by the newspapers of the day, including the person of Wilbur Rothar, is a complete fabrication; and 2) A man was inexplicably incarcerated under the name of "Rothar" for 24 years in mental institutions, without due process.

The huge discrepancies in press coverage, later magazine coverage, and the accounts of Mrs. Morrissey and GP Putnam's granddaughter in "Whistled Like a Bird" clearly indicate a fabrication that was altered with the passage of time.

Unfortunately, it appears that nobody has gotten their stories straight in the whole strange affair. In the end, we are left with one of the most bizarre sidebars to the Earhart disappearance.

Who was Wilbur Rothar? And what did he know that GP Putnam couldn't afford to let become public?

Chapter 10

Electra: The Sleight of Hand Begins

In 1936, the media reported that Purdue University had purchased a Lockheed Electra, Model 10E for Amelia Earhart's use as a "flying laboratory". Earhart, according to the media, immediately announced plans to use the flying laboratory for a full year of practical research into aviation. She was greatly interested in studying the effects on passengers and pilots of altitude, food and fatigue.

Initially, Earhart planned to complete her research first, then embark on her round-the-world flight. But as the plane's construction neared its end, she decided it would be better to complete her flight first, and then, with no pressure on her over a flight, attend to her research.

In the matter of the procurement of the Lockheed Electra, the truth was far more complex than reported in the media. For, in reality, funds had been covertly channeled to the university from unspecified sources. The government sleight of hand had begun. Placed in the newly created Amelia Earhart Fund for Aeronautical Research, the funds came to $80,000, equal to more than $1 million dollars in 2004 money.

As an early part of my investigation, I tried checking FAA records on the internet. Getting online one evening, I went to the FAA Records Center site to verify their current address. It was my intention to fly to Oklahoma City to personally hold and read the registration documents that Major Gervais reported discovering in their files.

Surprisingly, there was a caveat at the top of the page that due to security requirements, access to the public records room in the Oklahoma City Records Center had been suspended. Inquiries had to be made via either phone, fax, mail or the FAA's automated online server.

I immediately set about looking up, online, the two serial numbers Major Gervais had reported discovering in the FAA file: NR16020, Serial No. 1055; and N16020, Serial No. 1243. When I clicked on an inquiry link, a dialog box appeared, with just one field to fill out, the five digit registration number. There was no field for the prefix letter or letters.

Running the number 16020, I got a single entry for what was obviously the original Model 10E aircraft which had been purchased for Earhart in 1936. It was interesting to be sure, but not very useful. Oddly, the year that the number 16020 had been reserved was reported as 1987, which seemed strange at the time, since that Electra had been missing for 50 years. 1987 was also long after the model 12A had flown into the mountainside near Bicycle Lake.

Attempts to run queries using the manufacturer's serial number got a long list of aircraft, none of which was a Lockheed Electra. Not even Earhart's original plane was listed under Serial No. 1055. There simply was no way, online, to verify the existence of the other aircraft.

I therefore decided that the only way to make an adequate inquiry, under the current access restrictions, was to do it by mail. The next day, October 23, 2003, I drafted and mailed a letter to the FAA.

I very specifically requested copies of documents they had on Lockheed Electra, NR16020, Mfg. Ser. No. 1055, Model 10E and Lockheed Electra, N16020, Mfg. Ser. No. 1243, Model 12A. My purpose was to verify information reported by Joe Gervais and Joe Klaas about two Lockheed Electras being registered with Amelia Earhart's number, 16020. Gervais had reported in his book that he found in FAA records several documents reporting a 1940 transaction in which the Canadian government sold a Lockheed Electra Model 12A, Serial No. 1243, to the Charles Babb Co. in New York City.

Within just two weeks, a remarkably short time for the government, a large envelope full of material arrived in the mail from the FAA. It was a good news/bad news situation. The bad news was that, after a quick perusal of the documents, I realized that there were no document copies relating to the second Electra, Model 12A, Serial #1243.

Reviewing the initial letter to the FAA, I caught a serious typo made when I cited the registration number for the Model 12A. There was no choice but to resubmit a corrected request. Shortly thereafter, on 12-15-03, I drafted a second request, this time with the correct FAA number. I also inquired regarding the 1987 reservation of 16020, something I had left out of my initial letter.

The good news with the packet from the FAA was that it contained a small treasure trove of information on the Model 10E. It included various licenses and inspection reports. In addition, there were documents relating to the purchase of the plane and the last flight. There was even a copy of a memo reiterating radio traffic between Earhart and the ITASCA on the morning of July 2, 1937.

A sheet at the beginning of the stack of documents with Earhart's name and "Amelia Earhart File" on it made it seem that at least part of the materials were a standard package mailed to Earhart researchers.

Also among the materials received were copies of letters on G.P. Putnam's stationery and a copy of a letter on Lockheed Aircraft Co. letterhead. Most important, there was also a copy of the bill of sale for the Electra, and several documents relating to the CAA registration number of Earhart's plane. (See reproductions of these documents in Appendix III.)

It was the first document of interest, a copy of a two-page bill of sale for the original Model 10E Electra, that solved one small mystery. The document indicates that the aircraft was transferred to Earhart for the sum of $10.00.

After funds had been accumulated by Purdue, they were then used covertly to purchase the Lockheed Electra. Subsequently, as the newly discovered document indicates, the plane was transferred to Earhart in a token transaction for $10.00. As to the source of the funds, this is clarified by Randall Brink in his book "Lost Star". He quoted A.A. Potter, Dean Emeritus, of Purdue University Department of Engineering:

"Miss Earhart's plane was purchased for her in the interest of national defense. The money was channeled through two private individuals to the Purdue Research Foundation . . . Among her tasks was the development of direction finding equipment for the U.S. military".[9]

It has been speculated that the two private sources of funds for the aircraft were industrialists Floyd Odlum (a close friend of Earhart's) and Vincent Bendix.

In light of Mr. Potter's remarks, it would appear that the funds for the plane may have gone through the industrialists' hands, but, ultimately, came from another source. Private individuals don't make purchases "in the interest of national defense". Only governments do that.

The copies of the documents relating to Earhart's registration number helped to solve another mystery, one which was noted over 30 years ago by Joe Gervais during the research that resulted in "Amelia Earhart Lives". The crash of the Lockheed Electra near Bicycle Lake got Gervais thinking when he discovered that the craft carried Earhart's old CAA registration number. Gervais later developed evidence indicating at least one substitution of Earhart's ship.

---

[9] Brink, Randall, "Lost Star: The Search for Amelia Earhart" (NY: W.W. Norton Co., 1993, pg. 76.

LEGERDEMAIN

It was Klaas and Gervais who at least first discussed in print the remarkable variations in the appearance of Amelia Earhart's plane in various publicity photos. Trying to find an explanation for the CAA number on the craft that had crashed near Bicycle Lake, they began evaluating every photo they could find of Earhart's Lockheed. It promised to be an important clue to the Earhart mystery.

And indeed, if one sits down and looks at the many publicity photos of Ms. Earhart, it is immediately apparent that there are noticeable differences in the appearance of her ship. Differences which it is amazing that nobody else, in nearly 30 years, had noticed or remarked upon. These differences fall under two main categories: 1) Differences in CAA registration numbers; and 2) differences in physical appearance.

Interestingly, the CAA registration number on Amelia Earhart's plane is not always the same in the various publicity photos taken of her and her plane. In those photos, the CAA registration number is variously shown as NR16020, N16020 and R16020. (See Appendix II for a complete listing of the variations.)

The first document was a copy of a telegram sent by J. Carroll Cone, Assistant Director of Air Commerce, U.S. Bureau of Commerce, to G. P. Putnam. The telegram, dated September 25, 1936, authorized Earhart to change the number on her aircraft from R16020 to NR16020, and to display the telegram in the airship as authority for the change. Unfortunately, the telegram did not give the reason for the change in number.

The second item was a letter on G.P. Putnam's letterhead, acknowledging the September 25, 1936 memorandum, indicating the change would be made, and asking if another more permanent document would be needed for display in the plane.

These two documents partially solve the mystery of the multiple registration numbers visible in various photos of the Electra. Changes in the status of the craft by the FAA caused some of the changes in registration numbers.

But there is one discrepancy in CAA number to which this explanation does not apply.

Gervais managed to secure a photo of a Model 10E Electra with the number R16020 painted on it—the unique number assigned only to Amelia Earhart. That photo remained classified from 1937 until the 1960s, when Gervais was able to have a copy cleared and released to him. The craft in the photo was the XC-35, a top secret, experimental craft, which had the world's first pressurized cabin. That gave it much greater altitude capability. The XC-35 also had far more powerful engines than the "off the rack" Model 10E, thus giving it greater speed and range.

David K. Bowman

One could argue that the CAA number on the XC-35 had been routinely reassigned to Earhart. However the XC-35 was completed shortly before Earhart took off on her last flight, and therefore her number had already been assigned to her aircraft. Also the fact that the photo remained highly classified for so long argues that the presence of Earhart's number on the machine was considered a sensitive matter. Thus, there can probably be only one explanation for the XC-35 to be photographed bearing Earhart's number. Namely that the government had provided Earhart the use of the aircraft, or at least, was planning to provide her the use of the aircraft.

That is, *if* the photo Joe Gervais found was indeed of the XC-35. Was it?

Additionally, there are variances in actual physical appearance, which cannot be explained as changes in aircraft status by the CAA. For example, in some photos, the plane is shown with a loop style navigational antenna; in others the plane is shown with a navigational bubble, which is further aft on the fuselage than where the loop appears in other photos. Also, there are variations in wingtip navigational lights. Sometimes they are shown mounted at the ends of the wings, and at others, they are shown mounted on the front of the wings. Also, some photos show an elaborate paint job on the engine cowling and some show the cowling with no paint job. And as if that wasn't enough, in some photos, the door to the navigator's compartment has a window and in some, it is solid.

Even if one allows for one incorrect paint job and number variations caused by shifts in status by the CAA, the changes in physical appearance from photo to photo virtually cry out that, for whatever reason, Earhart's plane was switched at least once before she disappeared.

Another strong indication that at least one switch occurred is the fact that author Randall Brink was able to find and interview a Lockheed Aircraft Co. employee, who modified the belly of Earhart's plane so that Fairchild Aerial Survey cameras could be installed in the aircraft.

The employee was Lockheed airframe technician Robert T. Elliott: "This one was a Model 12," Elliott asserted. He candidly indicated that he did modifications to the Model 12A to allow installation of two (2) Fairchild aerial survey cameras. "That bit about repairing her crashed Model 10 was just a ruse," Elliott said at the end of the interview.

Brink also located Mr. Lloyd Royer, Earhart's unsuccessful suitor. He was employed at Lockheed during the 1930's, and had worked on Earhart's ship. He stated in a 1977 interview that the plane Earhart left in on her second attempt was different than the one used in the original attempt.

The above testimony would seem to nail it down that, following her failed March 1937 takeoff at Luke Field, Earhart's plane wasn't just repaired. It was replaced.

However, I happened to discover shortly before publication of this book, from Alex Mandel that several members of the Amelia Earhart Discussion Group had conducted a study of photos they could get of Earhart and her Electra. The approach was to identify metal sheets on the plane's skin, which they felt had unique shading and coloration. They then compared these with other photos. The result, according to Alex Mandel, was that the research team found these unique areas on all photos of Earhart's Electra. They concluded that the plane had not been switched.

Which is the correct answer? Is the Amelia Earhart discussion Group's study correct? Were the witnesses, Mr. Elliott and Mr. Royer, mistaken? This is another of those perplexing side mysteries surrounding the disappearance of Amelia Earhart.

On April 19, 2004, I was very surprised to receive a response to my December letter to the FAA. Again, it was a mixed bag.

On the up side, to the query regarding the reason for the 1987 date of reservation of FAA# 16020, the FAA indicated that the Earhart estate had requested the number be reserved in 1987 as a matter of course to prevent re-assignment of this historic number.

The negative side of the FAA letter was that in reply to a very specific request for copies of a series of specific FAA documents by FAA document numbers, the FAA indicated on their response form that they were "unable to identify the aircraft from the description given".

This seemed odd. Surely the much-vaunted FAA kept track of their records by document number? I took another stab at it, drafting another letter to the FAA, fully outlining the details of the 1940 and 1946 transactions, giving the model number and manufacturer's serial number of the aircraft. Just to be thorough, a request was inserted for copies of documents regarding the final sale of the aircraft in 1961. The document went out the next day, April 20, 2004.

Surprisingly, by the end of April 2004 a response from the FAA arrived in my mail. I was quite surprised by this sudden promptitude.

The response came on a standard FAA response form (AC 8050-69), typed at the bottom beneath a series of preprinted responses next to boxes. Dated April 28, 2004 it read:

"Our records show the aircraft Lockheed 12A, serial number 1243, formerly N16020, was destroyed in May 1965. The record is located in Federal Storage. We are requesting the file from storage and when received will review it for the requested information."

Aside from the fact that the reply was generated in a remarkably short time, there was one other unusual aspect to it. That was the fact that the crash date of the aircraft was cited to be May 1965. The crash at Bicycle

Lake occurred in the late fall of 1961. I put out a query to the Earhart Discussion Group, of which I am a member, I was told that delays in posting reports were common with the FAA. Apparently, the April 1965 destruction date was an error.

On August 16, 2004, eight months after my last query letter, I was astonished to receive another large envelope from the FAA. The envelope contained something like 200 document copies, some two-sided. These documents were all related to the second Electra, Model 12A, Serial No. 1243, the one Joe Gervais said came from Canada. (See Appendix VI for a summary of the major title-related document copies.)

What the second collection of document copies showed was that the second Electra couldn't have been connected to Amelia Earhart. Construction on it was completed in November 1937. Afterward, it was owned by both Lockheed and Western Air Express Corporation until 1940. Surprisingly, I discovered that Joe Gervais had gotten the transaction with the Canadian Department of National Defence turned around. For, in 1940, the Canadians bought the ship from Charles Babb Co. rather than selling the craft to this firm. The Babb Co. had bought the plane from Western Air Express immediately before selling it to the Canadians.

From then on, until 1946, the craft was in the possession of the Canadian government, apparently receiving little or no flying time. In 1946, Paul Mantz bought the plane after a series of three closely spaced transactions in which he had entered a deal to buy the craft from a company that did not even own it, Edward Ahr Co. Once this deal was completed, Algoma Air Co. purchased the craft from the Canadians and then sold the craft to Edward Ahr, who then consummated their deal with Mantz. (See Appendix VI.)

As strange as Paul Mantz' acquisition was of the Lockheed Electra, and despite his having kept the ship mysteriously out of sight during his ownership, there was no other connection between Amelia Earhart and the Electra than being owned by her former technical advisor.

In summary, the material in this chapter documents that even in its earliest stages, Earhart's round-the-world flight was the center of quiet machinations and manipulations; classic legerdemain, which has kept investigators confused for two-thirds of a century. The evidence in this chapter shows that after the crack-up at Luke Field, another craft might been substituted for the Model 10E Electra. What became of the Model 10E we do not know. The handling of Earhart's plane is like some strange shell game in which even the pea is substituted.

Chapter 11

More Sleight of Hand: Secret Modifications

In his account, Robert Myers brought up some discoveries he had made about clandestine changes that were made to the Lockheed Electra.

Mr. Myers said he made one discovery when he met a Mr. Max Clements in 1978. Mr. Clements was a researcher of the Earhart mystery, and the editor of Runway 26, an aviation-related magazine. It turned out that during Mr. Clements' investigation, he had discovered that Earhart's stop at Bandoeng was far from innocuous.

Max Clements found that a party was waiting for Earhart when she landed at Bathathia. She flew the party to Bandoeng, on another part of the same island, where, under the direction of Mr. F. O. Furman, they went to work on the Electra and installed powerful superchargers on the plane's Wasp Senior engines.

The installation occurred when it did because the superchargers had to be custom manufactured and weren't completed until late in the around the world flight. Earhart then flew the plane approximately 280 miles to Surabaya, to test the upgrade, returning to Bandoeng to complete a test flight of some 560 miles.

Mr. Clements promised to run a four issue, four part article in Runway 26 on Mr. Myers' experiences. But unfortunately, for some reason Mr. Myers never understood, only a small part of Mr. Myers' story was published, and the two men had a severe falling out. Mr. Myers never had any further dealings with Mr. Clements again.

Another of Mr. Myers' discoveries had to do with a remark that Fred Noonan made to Mr. Myers in 1937 at the Oakland Airport. It was just after the plane had arrived and Earhart reportedly had discovered it was a different plane than the one which had been crated up in Hawaii after the Luke Field crash.

Young Myers had been asking Noonan about the features of the new plane. Finally, Noonan told the young lad that every available void space in the aircraft had been filled with ping pong balls. Noonan told the puzzled youth that due to long expanses of ocean during parts of the flight, the ping pong balls had been installed to give the craft more buoyancy, should they have to ditch. Not long after, young Myers got a look at the ping pong balls, when one of the plane's gas tanks sprung a leak and the crew had to remove all the gas tanks to make the repair

Years later, in 1983, while lecturing at Iowa State University, Mr. Myers met a Mr. Rollo Christy. Mr. Christy told Myers he had worked on Earhart's plane in 1937, installing ping-pong balls in the aircraft. He also told Myers he had noted the presence of aerial surveillance cameras aboard the Electra.[10]

Some of the other telling things he said to Mr. Myers were, "The engines looked like they had been changed. They were bigger. That was also not an uncommon thing in those days. Mine Field, which is now an international airport, was pretty active. We did a lot of modification, but I doubt that her plane was completed when it left Inglewood.

"There were a lot of experimental things going on in those days, but one thing I was convinced of, I always swore up and down that if she went down in the water, there would be no way that plane would sink. Robert, so far you are the only person who has verified what I had thought".

---

[10] Myers, Robert, "Stand by to Die," Stand By to Die: The Disappearance, Rescue and Return of Amelia Earhart," (Pacific Grove, CA: Lighthouse Writers Guild, 1985, pg. 22-25)

Chapter 12

The Route

A critical issue in the disappearance of Amelia Earhart is the actual route she flew once she left Lae, New Guinea on July 2, 1937. Earhart's actual route has been at continual issue ever since her disappearance.

The official version says that she came within a few miles of Howland Island and then, running out of fuel, ditched in the sea, her ship going to the bottom almost immediately.

This version doesn't hold water (no pun intended) on at least two counts:

1. Evidence exists, from the accounts of Rollo Christy and Robert Myers, that Earhart's plane may have been outfitted with ping-pong balls in every available void space, in order to maximize buoyancy. As a result, Earhart's plane couldn't have sunk quickly. And in the event of a catastrophic breakup following a high speed crash into the sea, the wreck would have left a huge field of ping pong balls and other debris that the Navy couldn't have missed.

2. Also, there is credible evidence that Earhart continued to broadcast over her radio for as much as nine days after she ditched. The accounts of Earhart's own private network of ham operators in contemporary newspapers, as well as the account of Robert Myers' corroborate this scenario. It was only the U.S. Government that maintained, in 1937, in classic circular reasoning, that the ham operators couldn't be right because she had crashed nine days earlier and it wasn't possible. And since it wasn't possible, the ham operators had to be mistaken.

Investigators in the 1960s came up with another take on the problem. Two prominent ones, air force officer Paul Briand and newsman Fred Goerner felt, from their research, that Earhart had flown on an intelligence mission toward Saipan and ditched in Tanapag Harbor, to be picked up there by the Japanese. By the 1980s, some investigators returned somewhat to the government theory, and decided that Earhart had nearly reached Howland Island, but, worried that she had overshot, turned around and flew due Northwest—straight into Japanese Territory. There, they hypothesized, she ditched near Mili Atoll. A noted proponent of this hypothesis was Vincent Loomis. There is considerable evidence to show that one way or another the aviatrix did indeed end up near Mili Atoll in the Marshall Islands.

Robert Myers' report, which came out in his 1985 book[11], asserted that Earhart had ditched near Sydney Island, which is not far from Hull Island. He maintained that the aviatrix floated at sea for approximately nine days before being picked up by the Japanese. There is a possibility that Earhart may have been afloat somewhere for some days, although there is precious little evidence that she ditched near Sydney Island.

In 1987, James Donohue argued in his book[12] that Earhart had flown most of her published route, then diverted to the northeast once she passed Nauru Island, to make a pass over the Marshalls. After she had passed the Marshalls, according to Donahue, Earhart flew south east, deliberately bypassing Howland Island and heading for Hull Island.

Donahue had identified a section of the shore of Hull Island on which he said that Earhart had made a controlled, "wheels up" landing. He based this theory heavily on a radio study conducted by Paul Rafford. Surprisingly, in recent years, Rafford has changed his position given Donahue's book, and repudiated his earlier calculations. He has indicated he will outline his new position in an upcoming book, "Amelia Earhart's Radio".

Donahue also based his theory on reports from Pan American that their radio facilities had triangulated one of Earhart's broadcasts as coming from the area of Hull Island. This bit of information, to my knowledge, has not been controverted and, if valid, is the main evidence I have found that would support Robert Myers' theory.

Later, in the 1990s, Elgen and Marie Long in their book, "Amelia Earhart: The Mystery Solved," returned to the old official version of the route. The Longs maintained that the luckless aviatrix flew her published

---

[11] Myers, Chapter 5.
[12] Donahue, James, pg. 43-4.

route from Lae, almost reached Howland Island and then ditched within miles of her destination, sinking immediately without a trace.

Also in the 1990s, TIGHAR (The International Group for Historic Aircraft Recovery) launched expeditions to Nikumaroro Island (the current name for Gardner Island), on the theory that Earhart had overshot Howland Island and ditched on the shores of Gardner Island. (See chapter on TIGHAR.)

The Longs' hypothesis seems obsolete from the start in light of the material cited above, although TIGHAR's hypothesis was plausible. Both are untenable for several reasons:
1. There is substantial evidence to indicate a ditching in the Marshall Islands, possibly at Mili Atoll.
2. There is some evidence from various historical sources that Earhart may have continued to transmit messages over radio for as much as nine to ten days after she ditched.
3. No hard evidence has ever been uncovered on Nikumaroro that Earhart ever landed there.

Another aspect of Earhart's route was her "Plan B" or "Contingency Plan". According to Mary Lovell's biography of Earhart, the aviatrix was asked by Gene Vidal what her plan was if she failed to find Howland Island. Earhart replied that she would turn back and head for the Gilbert Islands, which were controlled by the British. Vincent Loomis was also aware of this when he wrote his book, cited above, and his theory undoubtedly was based on Earhart's reported Plan B.

The major reason Earhart selected the Gilberts is that options in other directions were limited. The Gilberts afforded a friendly government, many islands populated by friendly natives, easily available food and water, easy navigation with the sun at their back, and favorable tailwinds.

To the north, lie Japanese Mandate Islands that were fewer, more sparsely populated, with unknown food and water availability, and a hostile government.

To the south lie the Phoenix Islands, which were only lightly populated, with little food or water, and since they were fewer, they were harder to hit. If they are missed, the next stop would be Samoa or New Zealand.

And to the east, vast empty sea and distant Pitcairn Island.

Therefore, Earhart's only workable option was to turn west and make for the Gilbert Islands in the event of trouble.

Colonel Rollin Reineck brought up another possibility in his book. Reineck presented solid evidence that U.S. Army personnel stationed on Howland Island reported that they heard Earhart broadcast that she was turning north. Shortly afterward, Earhart radioed that she had run out of fuel, army personnel reported.

David K. Bowman

The strange thing is that ITASCA never reported hearing either of these transmissions. Could Earhart have been using a different frequency for some reason?

Yet another aspect of Earhart's route was the indications that she flew northeast in an arc over first the Caroline and then the Marshall Islands.

We are left with fairly persuasive evidence that an overflight of the Japanese Mandates, the Marshall Islands and possibly the Caroline Islands, did occur. It is also clear that Earhart had planned to turn west, in the event of trouble, and head for the Gilbert Islands, which was the only she could have done.

Was it really, as Reineck insisted, a secret Plan C, or did Earhart think she was flying towards the Gilbert Islands?

Chapter 13

The ITASCA

I was fortunate enough to be able to locate and talk extensively with David Bellarts, son of Chief Radioman Leo Bellarts, who was stationed on the ITASCA on July 2, 1937. Through David's kind auspices, I was able to obtain a lot of first hand information regarding arrangements and preparations on the ITASCA.

One of the things that David gave me access to was a duty assignment document, issued by Chief Bellarts' division officer, W. L. Sutter. According to the document, Chief Bellarts was assigned responsibility for monitoring all incoming transmissions from Earhart's plane. RM3 (Radioman Third Class) C. O'Hare was assigned to handle all other radio traffic received by the ITASCA. RM3 W.L. Galten was assigned to man the ship's direction finder when Earhart's craft came within 1000 miles of Howland Island. RM3 G.E. Thompson was the relief operator for all three of the above operators. RM2 (Radioman Second Class) Frank Cipriani was assigned to operate the direction finder on Howland Island.

Additionally, David gave me a copy of the logs of radio activity aboard the ITASCA relating to Earhart's flight from 1900 (7:00 p.m.) Howland Island Time, on July 1, 1937 through 1039 (10:39 a.m.) the next day, July 2, 1937. Over the years, the ship's radio log has been the subject of controversy, partially because its completeness and accuracy have been in doubt, and partially because of the heavy abbreviations in the log which are confusing. Initially, Chief Bellarts expected to be called to testify at an official inquiry, but that never happened. As a result, his copy of the log, the original, remained in the keeping of the Bellarts family from 1937 until 1974, when they donated it to the national archives. David maintains that the family copy has not been altered in any way.

Chief Bellarts, was on duty during the entire period that transmissions were received from Earhart's plane. That was from 0205 (2:05 a.m.) July 2 1937 until 0800 (8:00 a.m.). Chief Bellarts was not compelled to stand such a long watch, but insisted upon it out of a conscientious desire to personally be there.

Starting shortly before 7:00 p.m. July 1, 1937, according to the log, weak, unreadable signals began coming through. The following entries re Earhart's transmissions indicate what began to happen early on July 2 over the radio. To help the reader understand the entries, I have inserted a translation to plain language below each entry, in italics.

Chief Bellarts' radio log:

0400   EARHART HEARD FONE / WILL LISTEN ON HOUR AND HALF ON 3105-SEZ SHE BROADCAST WEATHER FONE 3105

Earhart heard via phone (voice). She said she will listen on the hour and half hour on 3105 kilocycles. She broadcast the weather by voice on 3105 kilocycles.

0453   HEARD EARHART (PART CLDY)

Heard Earhart, saying "Partly Cloudy".

0614   WANTS BEARING ON 3105 KCS // ON HOUR // WILL WHISTLE IN MIC

Earhart wants bearing on 3105 kilocycles on the hour. She will whistle in microphone.

0615   ABOUT TWO HUNDRED MILES OUT // APPX // WHISTLING // NW

About 200 miles out from Howland, approximately. Whistling in the microphone. Coming from the northwest.

0645   PSE TAKE BEARING ON US AND REPORT IN HALF HOUR

Please take a bearing on us and report in half an hour.

0646   I WILL MAKE A NOISE IN MIC – ABT 100 MILES OUT

I will make a noise in microphone. Am about 100 miles out from Howland Island.

0742   KHAQQ CLNG ITASCA WE MUST BE ON YOU BUT CANNOT SEE U BUT GAS IS RUNNING LOW BEEN UNABLE TO REACH YOU BY RADIO WE ARE FLYING AT A 1000 FT

KHAQQ calling ITASCA. We must be on you but cannot see you but gas is running low. Been unable to reach you by radio. We are flying at 1000.

0754   KHAQQ CLNG TASCA WE ARE CIRCLING BUT CANNOT HR U

KHAQQ calling ITASCA. We are circling but cannot hear you.

0800   KHAQQ CLNG ITASCA WE RECD UR SIGS BUT UNABLE TO GET A MINIMUM PSE. TAKE BEARING ON US AND ANS 3105 WID VOICE.

KHAQQ calling ITASCA. We received your signals but are unable to get a minimum. Please take a bearing on us and answer on 3105 kilocycles in voice.

0843   KHAQQ TO ITASCA. WE ARE ON THE LINE 157 337. WL REPT MSG. WE WL REPT THIS ON 6210 KCS WAIT. WE ARE RUNNING ON LINE LSNIN 6210 KCS.

KHAQQ to ITASCA. We are on the line 157-337. Will repeat message. We will repeat this on 6210 kilocycles wait. We are running on line, listening on 6210 kilocycles.

What follows next is the log compiled by RM3 Galten and RM3 O'Hare, who operated the onboard direction finder during the same period that Chief Bellarts was manning the radio. According to their entries, they monitored several frequencies, which included 500 kilocycles, 3205 kilocycles, 6210 kilocycles, and 7500 kilocycles. Another thing that was noted was that the men operating the direction finder also logged in messages they were receiving from another party or parties. The messages were generally inquiries as to the status of the flight, but it is not clear from the log whom they were from. I have included them where they are relevant to clarification, preceded by the letter "Q" for query. The reader should note that at the beginning of the log, RM3 Galten was on the watch. RM3 O'Hare relieved him at 0200. At 1033 hours RM3 Galten relieved O'Hare.

Direction finder log:

0400   Q: HAVE U ESTABLISHED CONTACT WITH PLANE YET

HRD HER BUT DNT KNW IF SHE HRS US YET. GAVE WX TO EARHART ON FONE 3105.

Q: Have you established contact with the plane yet?

Heard her but don't know if she hears us yet. Gave weather to Earhart by voice on 3105 kilocycles.

0440   Q: U HR EARHART ON 3105

YES BUT CAN'T MAKE HER OUT

Q: Do you hear Earhart on 3105 kilocycles?

Yes, but I can't make her out.

0455   EARHART BROKE IN ON FONE 3105 / NW???? UNREADABLE

Earhart broke in by voice on 3105 kilocycles. From the northwest? Signal unreadable.

0740   EARHART ON NW SEZ RUNNING OUT OF GAS ONLY ½ HR LEFT CANT HR US AT ALL. WE HR HER AND ARE SENDING ON 3105 ES 500 SAME TIME CONSTANTLY AND LISTENING IN FER ER FREQUENTLY.

Earhart on (?). She says she is running out of gas and has only a half hour left. She can't hear us at all. We hear her and are sending on 3105 kilocycles and 500 kilocycles at the same time constantly, and listening in for her frequently.

0857 AMELIA ON AGN AT 0800 SEZ HRG US ON 7.5 MEGS CA. STILL SENDING ON 7500 KCS TELLING HER TO CA ON 3105 AND SENDING OUT SIGS FER HER TO OBSERVE BEARINGS ON. MAINTAINING LISTENING WATCH 3105 KCS 7500 ES 500.

Amelia on again at 0800. She says she is hearing us on 7.5 megacycles calling. Still sending on 7500 kilocycles telling her to call on 3105 and sending out signals for her to observe bearings on. Maintaining a listening watch on 3105 kilocycles, 7500 kilocycles and 500 kilocycles.

In between the above entries were numerous remarks indicating:

- Transmissions by ITASCA in response to Earhart's received transmissions
- Transmissions by ITASCA in an attempt to contact and get an acknowledgment from Earhart
- Routine indications that no transmissions had been received from Earhart
- Lists of frequencies on which Bellarts was listening, which were 500 KCS, 3205 KCS and 6210 KCS

The foregoing material came from records held by the Bellarts family and seems to be above question as to its accuracy. David Bellarts maintains that his father was not a party to any alteration of this radio log. If the reader has any doubts, he can look at reproductions of the original logs in Appendix VII to see for himself. I would be happy to hear from any readers with input re the abbreviations used, which are indeed difficult to translate.

A major point of reference for investigators, the ITASCA radio log was initially only released to the media or cited in fragments. It wasn't until 1987 that a complete copy of the ITASCA radio log was finally released.

Although the above cited radio log material from Chief Bellarts' family seems totally reliable, the other two important records maintained by ITASCA, the ITASCA deck log and the Howland Island Detachment radio log have been found to be far more questionable.

In the August 2000[13] issue of Naval History Magazine, John P. Riley Jr. revealed in an extensive article that the deck log of the ITASCA had been partially falsified and that the Howland Island detachment radio log was almost completely bogus.

Riley concluded this from the accounts of surviving ITASCA crewmembers, particularly two radio operators, who were assigned to the Howland Island radio station. The two men were Yau Fai Lum and Ah Kin Leong.

Mr. Riley first wondered if the Howland log might be fictitious, when he showed Mr. Lum copies of the Howland log indicating that Lum had had stood radio watches on the direction finder on Earhart's frequency along with Radioman 2nd Class Frank Cipriani. Lum firmly maintained that those entries were completely fictitious and that he had not stood such watches, nor worked with Cipriani.

The other radio operators were Henry Lau (since deceased) and Ah Kim Leong. When asked what he knew about the situation, Leong declared in a 9-4-94 letter to Riley:

"No idea who wrote the false log. I stand no radio watch on Howland Island. Cipriani, Henry Lau and me was on the Coast Guard

Evidence, which Riley thought further, supports the thesis that the Howland log was falsified is the frequent misspelling of Lum's name as "Yat Fai Lum". "I should know how to spell my own name," Lum stated.

Overall, Riley felt that Lum and Leong were sincere and their stories merited more credence than the ITASCA documents, because both men had demonstrable integrity and neither man had any discernible motive to fabricate logs.

But here, we encounter another mystery in the Earhart disappearance: Chief Bellarts definitely made it clear to his family that Cipriani was not ordered aboard the ITASCA by CDR Thompson before the ship left to search for Earhart. Dave Bellarts is adamant about this point.

What's the answer?

There were two more pieces of information, which Dave Bellarts offered regarding what happened on July 2, 1937. According to his father, Chief Bellarts, there were no visibility problems over most of Howland Island that day, and also no problems with seagulls. This is contrary to most other theories and assertions that the main reason Earhart didn't or couldn't successfully land on Howland Island was excessive cloud cover

---

[13] Naval History Magazine, "Old Mystery, New Hypothesis," John P. Riley Jr., August 2000.

and/or sea gulls. At the moment, this contradiction remains unexplained, although this writer is inclined to believe the report of Chief Bellarts, and to believe that perhaps the glare of the morning sun on the sea near Howland Island caused Earhart's visibility difficulties.

Another piece of information that David Bellarts revealed to me, one which is probably being published for the first time, is regarding the new direction finder operated on Howland Island by radioman Cipriani. The direction finder, according to Chief Bellarts, was operated by rotating a wheel-like device from side to side in an oscillating pattern to try to determine the direction of a radio signal. The oscillating movement was necessary because there was a wire attached to the underside of the wheel, which would break if the wheel was rotated continuously in either direction.

Chief Bellarts later told his family that the direction finder had been disabled during its use on Howland Island, when the operator, Cipriani, rotated the wheel continuously without stopping. The damage wasn't discovered until a short while after the ITASCA had returned to Howland Island. At that time, the direction finder was brought aboard the ITASCA and Chief Bellarts dismantled and inspected the unit. He found the broken wire and also discovered that the device had failed during Earhart's flight due to overloading or running dry.

After repairing the unit, Chief Bellarts said nothing about it, according to David, because he knew that Cipriani hadn't received training to run the device and it would have served no purpose to get him into trouble.

As a result, it now appears that even when the direction finder's battery was good, the unit was inoperative. One has to conclude that due to a series of errors and oversights, the deck was stacked against Amelia Earhart.

David Bellarts provided yet another piece of striking information regarding the ITASCA. According to David, Chief Bellarts noticed during Earhart's attempted flight to Howland that documents were disappearing frequently from the "radio shack" [nautical slang for the compartment used on the ship for radio activity]. He reported this, and in short order the executive officer (second in command) of the ITASCA ordered Chief Bellarts to secure under lock and key all papers relating to Earhart's flight.

According to Dave Bellarts, his father suspected a particular ITASCA crewmember of the thefts, but was never able to secure incontrovertible proof.

Thus we have yet another couple of side mysteries in the disappearance of Amelia Earhart. Who was taking official documents from the ITASCA's radio shack—and why? And was Radioman Cipriani on Howland Island or the ITASCA?

Chapter 14

Miss Earhart Feared Forced Down at Sea

The following chapter attempts a presentation that to my knowledge no other writer has done in a book. That is to review in sequence the newspaper reports of the period of 2-17 July 1937 regarding Earhart's disappearance.

The idea was to see if any particular pattern might be present in media coverage of the disappearance. It turns out that there were, indeed, some definite patterns in the reporting, as well as some interesting things reported in the earliest stories. For one thing, most of the reports in the earlier news releases were either dismissed or contradicted by later ones. Below follows a brief survey of the newspaper coverage between July 2 and 17, 1937.

A typical headline early on July 2, 1937, in the Monessen [PA] Daily Independent reported "Coast Guard Tunes In Amelia On Hop To Tiny Ocean Island".[14] The sub-headline was "Racing for Pacific Islands".

The article reported in a routine, unworried manner that the aviatrix was "racing toward tiny Howland Island in the middle of the Pacific seas today," and went on to describe how the Coast Guard had made radio contact with the flyer at 6:18 a.m. that morning, Pacific Time. Earhart did not report her position, according to the story. One point of interest is the transmission referred to in the article as occurring at 6:18 a.m. Pacific time would have been approximately 4:18 Howland Island Time. And the only transmission in the ITASCA log which was close to that was logged in at <u>0400 (4:00 a.m.). The text</u> was:

---

[14] The Monessen Daily Independent, July 2, 1937.

EARHART HEARD FONE / WILL LISTEN ON HOUR AND HALF ON 3105-SEZ SHE BROADCAST WEATHER FONE 3105

Earhart heard via phone (voice). She said she will listen on the hour and half hour on 3105 kilocycles. She broadcast the weather by voice on 3105 kilocycles.

Was the above transmission the one referred to in the article? Otherwise, the tenor of this article was understandable, as Earhart had not yet been reported missing.

"Miss Earhart Feared Forced Down at Sea," the Reno Evening Gazette announced later on July 2nd, in a bold headline of medium size in the middle of the front page.[15] Now there was a note of alarm in the coverage. The story reported that Earhart and Noonan were believed forced down near Howland Island that morning. Earhart, according to the report, had earlier radioed that she had about half an hour of fuel left and that no land was in sight. The article mentioned that the Lae-Howland leg of the flight was a potential segment of a commercial route from Australia to America. It concluded with a brief recap of Earhart's itinerary, ending with her original ETA of 8:30 a.m. at Howland Island.

The aviatrix' disappearance commanded the front pages on July 3. "Battleship Joins Search for Amelia," the Charleston Daily Mail for that date declared in huge print at the top of the page. The subheadline read "Famed Flyer, Aide Believed Afloat in Sea". In fine print it was clear that the USS COLORADO may have been ordered into the search, but that the vessel had only just reached Honolulu. Additionally, according to the article, naval units from the University of California and University of Washington were aboard the COLORADO. Since the only military units attached to colleges and universities are ROTC units, we can be sure that the personnel from those two universities were ROTC cadets.

"Weak Radio Signal Starts Hunt in Island Area,"[16] another subheadline announced. This evidently referred to references in two other small articles on the front page to SOS signals received by ham operators in Los Angeles. The operators identified were Walter McMenamy and Carl Pierson, members of Earhart's radio network, who were reported to have heard

---

[15] Reno Evening Gazette, July 2, 1937.
[16] Charleston Daily Mail, Saturday, July 3, 1937.

a voice identified as Amelia Earhart calling "SOS, SOS, SOS, KHAQQ, SOS, SOS, SOS". This was shortly before 5:30 a.m. Pacific Time.

Still another article in the Charleston Mail for Jul 3, 1937 told the public that GP Putnam had just finished conferring with Paul Mantz over the phone and both men were convinced that Earhart had made a safe landing on a small island somewhere off her course.

By July 3, it was quite clear that a steadily expanding and frantic search was in progress for Earhart.

The front pages projected strong apprehension on July 4, with the headline "Storms Endanger Earhart," in huge bold print at the top of the front page.[17] The story reported that storms had turned back a navy seaplane which had been speeding towards the Howland Island area from Honolulu. This left, according to the newspaper, only the ITASCA to carry on the search for the missing aviatrix. The article recapped news releases of previous days and indicated that no new radio signals had been received from Earhart.

Headlines the next day were even more distressing: "Earhart's Airplane Reported Sinking" the Charleston [WV] Daily Mail declared in large print at the top of the front page.[18] This article was particularly surprising to me because the story reports that Earhart had made a radio transmission after ditching at sea, and most researchers since 1937 have dismissed the idea of Earhart making transmissions after landing in water.

The message Earhart was reported to have sent was "281 North Howland call KHAQQ beyond north don't hold with us much longer shut off". Admittedly garbled, the message was reportedly intercepted by three different navy operators, who pieced the communication together afterward.

Another paper, the Newark Advocate, had a hopeful note later that same evening[19] with the headline, "Earhart Signals Revive Hope For Early Rescue". The centerpiece of the article was the news that station KGMB had broadcast a message to the missing fliers earlier that day. The message ran:

"Earhart plane. Come in on 3105 kilocycles. Use key or voice. Key preferably. Put transmitter on one minute then take off and turn on four times. Follow these instructions immediately after this broadcast".

---

[17] Lima Peru News, Sunday, July 4 1937.
[18] Charleston Daily Mail, July 5, 1937.
[19] The Newark Advocate, July 5, 1937

Almost immediately after the broadcast, listening operators heard four distinct breaks over the radio.

Another broadcast was reported by station KGU: "If on water, send eight dashes. If on land remain silent". A short while later, coast guard and army radio operators reported hearing a series of dashes over the air. How many dashes had been received was not clearly indicated. However the inference is that there were eight. If bonafide, this would indicate that Earhart had ditched at sea.

The tenor of these reports, said the newspaper, was that hopes had been raised for the rescue of the missing aviatrix.

On July 6, yet another message was reported. "Radio Message Gave Position of Earhart Plane" the headline of the Chillicothe Constitution-Tribune[20] announced. Charles McGill, an amateur radio operator in Oakland, CA was reported to have picked up the following message on 3105 KCs:

"NRUI [the call sign for the ITASCA] – KHAQQ – SOS SOS SOS – KHAQQ – 281 North Howland. Can not hold out much longer. Drifting slowly southwest. We are above water. Motor sinking in water. Very wet".

The story went on to indicate that GP Putnam, Earhart's husband placed a guarded credibility in the message, and described Putnam's reasons: internal consistency and the timing of the message.

"No Word of Lost Fliers is Received," the July 7, 1937 issue of the Helena Independent reported.[21] Of significance, the article stated that the focus of the search had shifted from the area north and northwest of Howland, to Winslow Reef 175 miles east of Howland.

Another fact of significance reported in the Helena paper was the fact that full command of the search effort had been given to Admiral Orin G. Murfin, Commandant of the 14th Naval District in Honolulu.

The next day, July 8, 1937, the Reno Evening Gazette reported "Hunt for Amelia Earhart is Concentrated Today in Group of Phoenix Islands".[22] If this report was accurate, the navy had shifted its focus both to the east and the south, as the Phoenix group is located south of Howland Island. "Planes Continue to Scan Waters of South Pacific" a sub headline continued, and the article described the activities of the navy planes searching the area. The bottom line for the article, however, was that there were no new

---

[20] Chillicothe Constitution-Tribune, July 6, 1937.
[21] Helena Independent, July 7, 1937.
[22] Reno Evening Gazette, July 8, 1937.

developments in the search. But of note, was a mention that the Carrier Lexington had not yet arrived, but was expected in Honolulu at about 8:00 p.m. eastern time (which would be about 3:00 p.m. Honolulu time).

"No Clue Yet to Plane as Hunt Widens" was the word in the wire service story on July 9, 1937.[23] The story was largely a rehash of previous developments, but also indicated the most recent movements of navy ships, including the fact that the Lexington was underway for the Howland area with fuel for the other search ships.

But the end of the article contained some interesting information. A new message was reported to have been received from Earhart. Arthur Monsees, a ham operator in the San Francisco area was reported to have received a short message, "East Howland – lights tonight – must hurry – can't hold," on 6250 KC. Authorities were reported to be giving credibility to the message, although they felt that Monsees had erred in his frequency calibration, as Earhart's frequency was 6210 KC.

Later that same day, another press release sounded even more bleak: "Hope Ebbs For Lives of Flyers Lost in Pacific".[24] The article reiterated most of the contents of the earlier release, concluding with the observation that GP Putnam continued to believe that some of the messages reported earlier in the search were bonafide.

"Earhart Hope Wanes" the papers announced in large print on July 10th. The sub headline was "Navy Plans Final Check of Huge Area"[25]. According to the news release, although naval forces in the area of Howland had found no trace of the two flyers, the Carrier Lexington with a compliment of 63 planes still offered some hope of success in the search.

The Lexington was reported to still be enroute to the Howland Island area.

The release concluded that no new messages of any sort had been received or purported to have been received in the previous 60 hours. GP Putnam was reported to still believe that several of the previous messages were authentic. And, as had been reported in previous releases, the government was reported as skeptical of all the previous messages. Also, GP had been reported as believing that the authentic messages from Earhart had been sent while the plane was on dry land.

---

[23] Helena Daily Independent, July 9, 1937.
[24] Edwardsville Intelligencer, July 9, 1937.
[25] The Lima News, July, 10, 1937.

On July 11, the Lima News reported "New Areas Scanned With Slight Hope of Finding Miss Earhart"[26]. The Carrier Lexington was reported to be nearly in the search area, with all aircraft fueled and ready to launch.

he news of July 12 was the first harbinger of the failure of the search. "U.S.S. Colorado Withdraws From Earhart Search" the papers reported somberly.[27] The withdrawal of the Colorado was due to the impending arrival of the Lexington, which was reported to be steaming at half speed to conserve fuel. It was reported that the Lexington would not be expected to be involved in the search until the next day (Tuesday) or the day after that.

"Hope Not Abandoned," the Indiana Evening Gazette for July 13, 1937 told the world. This small headline was just above a photo of Earhart and Noonan taken shortly before their disappearance. The brief article reported that the "greatest war air force ever assembled for a peacetime mission was poised today to skim along the mid-Pacific equator" to search for Amelia Earhart. Evidently, the Lexington had been further delayed by a tropical storm and had close to reaching the Howland Island area. The article concluded with a description of Howland Island and plans for the next area to be searched—an area 60 miles wide and 600 miles long "extending north and south from Howland Island".

By the next day, the Earhart disappearance had receded a bit in the news. On page five of one newspaper[28] there was a small article, headed, "Continue Search for Miss Earhart". It reported that the search for Earhart was continuing, although bad weather had stopped the previous day's sorties by sometime in the midday. The search was reported as "fruitless" and that hope was "at the vanishing point".

One news release for July 15, was even smaller than the July 14th article. "Navy Continues Search For Earhart" was heading of the two paragraph article.[29] The piece briefly reported that the search was still continuing for the missing flyers and that chances of finding her were "placed at one in a million".

Earhart Hunt Nearing Finish," was the heading of the small article on page four of the Charleston Daily Mail for July 16, 1937. The navy was reported to be determinedly struggling through terrific heat and heavy tropical squalls in their attempt to find the lost flyers. The article concluded with the remark that

---

[26] Lima News, July 11, 1937.
[27] Chillicothe Constitution-Tribune, July 12, 1937.
[28] Chillicothe Constitution-Tribune, July 14, 1937.
[29] Monessen Daily Independent, Monessen, PA July 15, 1937.

the navy would probably end the search on Saturday July 17th. But, as it turned out, the search was finally discontinued on July 18th.

Reviewing the articles, I noticed that the early articles reported that Earhart had landed at sea and was broadcasting radio messages from her plane. They even gave the text of the messages. Within a few days, a new change was introduced when GP was reported to have heard messages from Earhart in which he could hear the roar of engines in the background. GP was reported to be convinced—at the time—that Earhart had landed on dry land and had spent two days broadcasting SOS messages, using the plane's engines to generate power.

As reportage progressed, the tenor quickly and completely reversed to the conclusion that the aviatrix had simply run out of fuel near Howland and had plunged into the sea, perishing immediately. Virtually all the messages reported during early coverage of Earhart's disappearance were dismissed by the government as being "unconfirmed".

Interestingly, the above pattern has occurred in other situations where something controversial is reported to have happened and then a short time later, the tenor of the news releases changes quickly. In other words, quite frequently, early news reportage of an important event is usually the most accurate. Later reports, in short, are sometimes possibly "damage control".

The fact that Earhart was reported by the media initially as having ditched at sea and having made multiple broadcasts, contradicts later assertions that she couldn't have made a broadcast after ditching at sea. Surprisingly, careful scrutiny of the facts does not find absolutely incontrovertible evidence that rules out radio usage while floating in water.

In summary, the newspaper coverage initially indicated Earhart ditched at sea and broadcast SOS messages from her plane. Despite later indications that she might have landed on dry land, the indication was still that Earhart had not perished immediately. After the first week or so, for some reason, the reportage then changed direction. For the rest of the period of the search, the thrust of reportage was that the aviatrix had simply been lost at sea near Howland Island.

Why had all the published radio messages from Earhart been so completely dismissed by the establishment? And why would "damage control" have been needed in the media reportage of the Earhart disappearance?

Chapter 15

Interception

Just which Japanese naval vessels were in the area where Amelia Earhart disappeared, which ones might have seen or forced down the Electra, and which one might have actually retrieved Earhart's craft are key issues in the Earhart mystery.

Not every researcher subscribes, of course, to the notion of Earhart having been picked up by the Japanese, but there is enough evidence for that proposition and enough controversy as to how it may have happened, that it is constructive to try to deal with this issue.

Several researchers, including Joe Gervais and Randall Brink assert that on the day that Amelia Earhart went missing, the carrier AKAGI and its related task force were engaged in maneuvers near Jaluit Atoll. They posit that Earhart was spotted in Japanese airspace by the carrier, which then launched planes to force down the Electra.

Both Gervais and Brink cite the testimony of Mr. Fujie Firmosa, a former Japanese who claimed to have been assigned to the Carrier AKAGI. Mr. Firmosa recounted that he was one of a number pilots scrambled to intercept Earhart's plane when it was detected in Japanese airspace. According to Firmosa, after encountering Earhart's aircraft, he made a second approach to the plane, in what he described as a "firing pass." He stated that after this pass, during which he fired a burst of bullets at the Electra, the aircraft then crash landed near Mili Atoll.

This account seems reasonable until one checks and finds that the Carrier AKAGI was in Sasebo Shipyard for a major refit from 1935 until 1938. I checked multiple sources, as well as checking with Russian scientist Dr. Alex Mandel, a fellow member of the Amelia Earhart Discussion Group who is a published authority on naval history. What I found was that the AKAGI couldn't have been near Jaluit in July 1937.

Alex was quite certain that the ship cited by witnesses in the area of Jaluit in July 1937 could not be the AKAGI. According to him, the AKAGI was not only in Sasebo in July 1937, but was also extensively dismantled and incapable of operations.

But Alex did suggest that another Japanese aircraft carrier was active at that time and could have been in the area. This was the Carrier KAGA. Unfortunately, there is no clear information, Alex says, about the exact location of KAGA in July 1937.

I also queried Joe Klaas by email, and he acknowledged the conflict between Japanese records and the testimony of various witnesses regarding the AKAGI. He didn't indicate the reason for the conflict, but inferred that possibly the Japanese records may have been falsified to obscure the actual locations of some of their ships. Apparently, Gervais and others feel that the eyewitness testimony is more reliable.

And indeed, it may be possible that the Japanese altered some records to obfuscate some of their ship movements. However, evidence to the contrary, particularly from naval history sources like Dr. Alex Mandel, seems more persuasive to this writer.

Therefore, if there was a Japanese naval task force in the area of Jaluit, which could have spotted and forced down Earhart's plane, it most certainly was not one associated with the AKAGI. Quite possibly, per Alex Mandel, it was the KAGA, a ship with a name very similar to the AKAGI, which could have been confused with that of the AKAGI by researchers and witnesses alike. Or possibly garbled by translators for the researchers or witnesses.

The other main question regarding the possible retrieval of Earhart, Noonan and their craft by the Japanese is what might have happened after the Electra went down near Mili Atoll. This isn't as controversial as the matter of the AKAGI, and it appears from eyewitness testimony that Earhart's ship was first found by a Japanese fishing boat.

After this was reported by the fishing boat captain, a Japanese ship was reportedly dispatched to retrieve the aircraft. This reportedly occurred nine or ten days after Earhart ditched.

For years it was felt that the Japanese Seaplane Tender KAMUI (generally misspelled in books and articles as "KAMOI") was the ship which might have picked up Earhart, Noonan and their plane. However, no evidence has been found that KAMUI was in the area of Mili Atoll.

Another ship, the coal tender KOSYU (commonly phonetically misspelled as KOSHU) has more recently been identified by researchers

LEGERDEMAIN

as the most likely vessel. Testimony has surfaced from residents of the Marshall Islands which seems to corroborate that indeed the KOSYU was sent to Mili Atoll to pick up Earhart's craft some nine days after she went missing. (See "The Eyewitnesses")

The available eyewitness evidence shows that after the Japanese fishing boat picked up Earhart, it took her to Mili Island. There, surprisingly, the authorities may have allowed her to make at least one radio transmission to the U.S. government. They may have initially thought Earhart and her navigator to be innocent lost aviators. Shortly, they contacted Japanese headquarters in Tokyo and were told to hold onto their charges until a ship arrived to remove them from the island. Thereafter, a Japanese naval vessel, quite possibly the KOSYU, retrieved the downed Electra, and proceeded to the island of Mili to pick up the two aviators being held by local authorities.

If the above is correct, then just as the huge U.S. naval search force reached the Central Pacific some nine to ten days after Earhart disappeared, the aviatrix and her navigator were already in Japanese custody.

The resulting search was thus moot, with the world-famous flyer behind their reach.

Chapter 16

What will they do to us?

Joe Klaas' and Joe Gervais' book, "Amelia Earhart Lives," unwittingly stimulated a further, and possibly important development, which may tend to confirm some of Klaas' and Gervais' contentions.

I discovered the following account when I obtained a copy of a little known 1985 book entitled "Stand By to Die: The Disappearance, Rescue and Return of Amelia Earhart," written by Robert H. Myers.

Mr. Myers' account is problematic, however, because portions of it conflict with a number of points considered to be known facts. Nevertheless, Mr. Myers' account is so unusual and remarkable that it deserves attention in this report.

One thing I noted is that the first part of the book's title harkens back to the original title of the screenplay of a film. Entitled "Stand By to Die," the film was a thinly fictionalized account of the disappearance of Amelia Earhart. During the making of the film, which ultimately was released as "Flight for Freedom," GP Putnam, who had actually been in on the development of the screenplay from the beginning, launched or threatened to launch a lawsuit against the studio for inferring the film had to do with his wife. This was apparently just a publicity stunt, which paid off handsomely, as the film is still remembered over 60 years after its 1943 release. The film must have made quite an impression, as Mr. Myers' used its original name for the title of his book.

A California resident then in his late forties, Mr. Myers happened in 1970 to see a news film of Irene Bolam denying that she was Amelia Earhart, in a news conference held after she filed suit against Joe Klaas and Joe Gervais.

In Myers' book, he described how he had met and befriended the world famous aviatrix, Amelia Earhart, in the spring of 1937 at the Oakland Airport, when he was just fifteen years old. The youngster got to know the flyer quite well and was extremely familiar with her appearance, voice and mannerisms.

When he saw the news film, Myers was immediately certain that he was looking at his old friend, Amelia Earhart. He was stunned. Up until then, he had assumed Earhart had perished at the hands of the Japanese. Enormously relieved, Myers resolved to further investigate Earhart's disappearance himself. Eventually, after making his own investigation and speaking to many people, including Irene Bolam herself in multiple conversations lasting for nearly four years, Myers told his story in his 1985 book.

The entire book was told not as Amelia Earhart's story but as Robert Myers' story, and quite effectively, as the account was told in the completely honest, forthcoming manner of an ordinary, innocent young man who had had some extraordinary experiences.

In his book, Myers described how, during his frequent visits at the age of 15 to Bay Farm Airport, as the Oakland Airport was known in 1937, he met Amelia Earhart. Earhart and the boy became fast friends, and Myers was soon a regular at the airport, frequently being allowed to be present during conversations between Earhart, Noonan, Mantz, Putnam and other participants in Earhart's last flight.

Young Myers was able to closely observe Earhart and her colleagues as people, as well as their day to day activities at the airport. At the time, all activities at the airport not relating to Earhart's flight had been stopped and the whole facility was devoted to preparation for Earhart's world flight.

Myers' account provided a heretofore unknown and unavailable inside glimpse into both preparations for the world flight and also the personalities of the participants. Most of the people involved were remarkably affable, and Earhart, in particular, endeared herself to young Robert Myers with her warmth and friendliness.

One thing young Myers frequently observed was government representatives coming and going at the airport. Interestingly, whenever government agents arrived, GP usually sent Earhart to the coffee shop while he dealt with the agents. From time to time, Myers reported, Earhart and Noonan sat in the coffee shop fuming about GP's keeping them "out of the loop".

After Earhart returned from Honolulu following her crash at Luke Field on her first attempt at the round-the-world flight, young Myers immediately noticed that the Lockheed Electra Earhart had flown to

Honolulu was placed in the middle of an airstrip and covered with a tarp. Every day mechanics went out and worked on the plane, under the tarp, under conditions of great heat and discomfort, literally stripping every item of equipment from the plane.

A short while later, the Electra was taken away and then ostensibly returned to the airport. Myers immediately realized it was a different ship, recounting that Earhart also noticed the difference. She had evidently not been told about the switch, and was visibly displeased.

Several days before Earhart flew out of the airport bound for Miami on her second round-the-world attempt, a scene occurred that was to have repercussions on young Myers' life for years to come.

As Myers entered the coffee shop that day, Earhart and Noonan were in the middle of an intense conversation regarding GP.

Earhart said, "I don't care what they tell you, Fred. There is something going on and they are not telling you and they are not telling me. He [Putnam] won't tell me a thing and I know he is up to something. He won't even talk to me anymore. He doesn't tell me anything. Do you know who all of these people are?" (Referring to the various government people visiting the airport).

Noonan said he didn't and that he knew no more than she did. At that point, Earhart offered Noonan the chance to back out of the flight. Noonan declined, and then as both of them noticed young Myers in the room, they dropped their voices.

Uncomfortable at having walked into a private conversation, young Myers started to the leave the coffee shop. Earhart stopped him, placing her hands on his shoulders.

"Robert, I want you to make a promise to me. Can you keep a promise?" Robert said he would.

"You know who I am and what I do and I want you to promise me to do something. I am on a very secret and dangerous mission and I want you to tell if you hear that anything happens to Fred or me. Tell your mother or someone else. Will you promise me that?"

At that, Noonan hopped off his stool and said sharply, "What the hell did you tell him that for? You told me not to tell anyone!"

Remonstrating with Noonan over his language, Earhart continued, "Robert is a big boy, and he will tell someone if anything happens. Do you promise, Robert?"

Robert nervously promised he would. When Earhart smiled at him and patted his shoulder, he felt a little better and promptly left the shop, inadvertently leaving through the front door. He usually left via the back door and through the hangars, but was distracted by the intense scene in the coffee shop.

Not far from the coffee shop, Myers encountered Putnam, who shouted at him, "What did they tell you in there?" There followed an ugly scene according to Myers, after which Myers rapidly walked away.

Since it was late and there was no way for young Myers to get a ride home, he began walking. After a while, a car came along behind Myers and he thought "At last a ride!" But when he turned around he saw, incredibly, GP Putnam's 1937 Hudson bearing down on him. Jumping quickly aside to avoid getting hit, Myers landed in a ditch. The Hudson came to a screeching halt and before anything more could happen, another car drove by and gave the frightened young boy a ride home. Myers later reminisced in his book that he clearly recognized GP Putnam at the wheel of the Hudson.

This incident, if true, and there is no indication that Mr. Myers was not truthful in his account of it, is a disturbing one, as nothing anywhere else in the material about Earhart shows GP in such a dark light. The incident underscores the fact that Earhart was indeed involved in a secret mission for the government. It also makes it clear from Mr. Myers' account that the relationship between Earhart and Putnam seemed to be in serious trouble just before Earhart disappeared. Moreover, the incident, if true, proves Earhart was on a secret mission, but did not know all of the details. The author leaves it to the reader to draw his own conclusions about GP from the foregoing account.

Several days later, Earhart took off to the east on her around-the-world flight.

But more intriguing experiences were in store for Mr. Myers. His father had purchased a new Philco radio, which could receive short wave broadcasts, and installed a 60 foot copper mesh antenna outside the house. This gave the family radio considerable range, allowing them to enjoy all sorts of radio programs and broadcasts.

In the evening of July 2, 1937, young Myers couldn't sleep and got up to listen to the radio. He wasn't supposed to listen to the radio without his father present, but fudged a bit on this rule. Surprisingly, when he turned on the radio, he heard the voices of radio operators on the Coast Guard ships ITASCA and ONTARIO.

Myers became entranced and ended up staying up for the rest of the night following not only the radio conversations of the Coast Guard operators but those of Amelia Earhart as well. Myers was amazed that he was able to hear Earhart's voice over his radio, and followed the Lae to

Howland leg from the moment of takeoff. He kept detailed records of her transmissions and the times they occurred.

Myers recounted that these transmissions continued amazingly for nine days, during which the rest of his family members were frequently present and heard the transmissions along with him.

In his book, "Stand By to Die," Myers spent an entire chapter detailing Earhart's transmissions during the Lae to Howland leg of her flight, and an entire chapter on her transmissions over the nine day period after she ditched.

What was so amazing was the depth of detail in the transmissions and the fact that Earhart repeatedly transmitted her location over the entire nine day period, until finally a Japanese vessel boarded their floating plane and took them into custody.

One of the most amazing transmissions Myers reported was one in which Earhart spoke with Harry Balfour, of the New Guinea Airways, over the radio. Although it's widely known that Earhart and Balfour maintained radio contact for some eight hours after she took off from Lae, the exact contents of the transmissions aren't.

In the transmission, Myers reported that Earhart told Balfour that Fred Noonan had passed a sealed letter to her on the fishing pole they had installed over the extra fuel tanks. She said that she would go off the air for a short while to read the letter.

Fifteen minutes later, Earhart got back on the radio to Balfour and informed him that she had been given secret instructions to turn northeast toward Truk. Evidently she was not aware of the plans for her to divert to the north over Japanese territory.

Earhart asked Balfour if he was aware of the secret instructions. He replied that he wasn't and that he was only told to transmit weather reports to her until she was about 400 miles from Lae.

They want me to fly directly north toward Truk Island by way of the Admiralty Islands, first of all. If you can hear me, the Big Arch, did you know that? The Admiralties! The Admiralties! Oh my goodness. They should have told you! They should have told you about this! I am not sure if I want to do this! It is different. Who am I going to talk to? Who is going to give me the radio reports?"

This assertion suggests that Earhart thought that the Lae-Howland leg of the flight would be flown along the published route. Some skeptics have dismissed the idea of Earhart's having flown over dangerous Japanese air space on a military type mission, citing Earhart's well known tendencies toward pacifism and dislike of anything related to war. This transmission explains how and why that could happen: She was tricked into it.

Up until then, Earhart had merely taken an occasional surreptitious photo of an airport or airstrip here and there, particularly in Africa. It may be that that was the only part of the secret mission on which she had been briefed. And for obvious reasons. She would never have agreed to venture into such a perilous situation if she had been asked, both because of the danger and the fact that it would have represented to her a clearly military-type flight. And it's also hard to believe that she wasn't aware of the extremely hostile posture of the Japanese at the time, as she was an extremely intelligent and well-informed woman.

Near the end of the Lae-Howland leg, Robert Myers reported that she had reached Howland Island, but had trouble seeing clearly to land. There is a contradiction here, as David Bellarts has indicated that his father maintained that there were no visibility problems at Howland Island. Strangely there were no acknowledgements from ITASCA to Earhart's transmissions.

Young Myers then reported that he heard the voice of a ham radio operator come on over the frequency, desperately trying to tell the Coast Guard that he knew Earhart's location. At that point, Myers was thunderstruck when he heard the voice of the commanding officer of the ITASCA, Captain Thompson come on over the air and tell the operator, "Get the hell off the frequency".

Myers was amazed at the whole series of incidents because it confirmed that other people were hearing the same transmissions he was, but that the Coast Guard wasn't interested in being told about them. At least twice after that, ham radio operators came on over the frequency to try to advise the Coast Guard of the transmissions they had heard. On each occasion, the Coast Guard sternly chastised the ham operators and did not seem to show an interest in receiving the information.

Earhart then indicated that Noonan had recommended she try for Hull Island. Shortly afterward, she broadcast a more complete location than has ever been published in the media except for Mr. Myers' account:

As she approached Hull Island, Earhart radioed that, "We are only six miles out. It is getting close. We are flying north and south 157-337. Our reference point is Baker Island. Can you hear me? Use the rose of the compass and draw a line through it. I hope you hear me. Use the rose of the compass from 336-157. Draw a straight line and that is where we are!"

A moment later, Earhart reported over the radio that three Japanese fighter planes were approaching her ship; she even described the "rising sun" insignia on them. Within moments, she saw one above her plane and one near either wingtip. Abruptly, they began firing their machine guns in short bursts.

Just then, Earhart spotted a large ship and its escorts, along with a fishing vessel.

"There's a fishing vessel, a small battleship and a large one. It is so huge. Fifty times bigger than you! It is unbelievable! You won't believe this. I have never seen a ship as big as this one".

A second later, Earhart received a note from Noonan to land on Sydney Island, less than a hundred miles east. She changed course.

Within eight miles of Sidney, Earhart reported that one engine had gone out and minutes later she ditched. Young Robert Myers could hear the sound of her plane hitting the water.

Overall, by Mr. Myers' account, there was nothing ambiguous about Earhart's circumstances or location in her transmissions. It was always clear to him where she was and what was happening. Her ditching not far from Hull Island, coincides remarkably with the determinations of James Donahue, to be discussed later in this book.

After Earhart had ditched, she made some remarkable transmissions that cast a whole different light on her disappearance. After Earhart and Noonan had been adrift for five or six days, Earhart and Noonan vented over the radio their frustration about GP's secrecy and their concerns that he was the reason no help had arrived. They speculated that the widely published rumors of a love affair between them had upset GP, and Noonan told no one in particular over the radio, "Whatever she says, you can believe her. He's doing this. I don't know why he is doing this. There is not a love affair or anything like that".

The following is one transmission that conveys the tenor of a number of Earhart's subsequent transmissions:

"It's so hot. I can't stand it much longer. It's unbearable. Please, George, don't do this to us. You can hear me. I know you're listening for me on their radio. If you hear me, please tell them where we are. You can. Please tell them where we are like you promised me you would do".

Yet another was just as startling:

"Oh please, hear me now, Captain Black. [Captain Richard Black, who provided the secret direction finder equipment for Earhart's flight.] I know you are there and you know my voice. Please, if you hear me, make my husband tell you where we are. He knows about this and he knows where we are. Make him tell you about this. Please, Mr. Black, or anyone who hears me. What will they do to us?"

Could Earhart have been transmitting on a different frequency than that the ITASCA was monitoring? As reported earlier in this book, the FBI had sworn the ham radio operators to secrecy. Did all transmissions go first to GP? And was GP then supposed to pass on information he received through the radio network to the Coast Guard via FBI agents? Did GP actually suppress some transmissions from Earhart?

The whole situation seems outrageous, except that Mr. Myers' account appears to be so straightforward and honest, filtered through the memories of an innocent fifteen year old boy.

If Earhart was not delirious when she made those pleas over the radio, the implications are disturbing to say the least.

Of course, a major problem with Robert Myers' account is that no other radio operators reported hearing the transmissions that Myers reported. Why *didn't* someone else hear any of those transmissions?

One of the things about Mr. Myers' account that lends it great credence, is the simplistic political outlook he reported at the time of the Earhart disappearance and his reaction to having heard over the radio of her being picked up by the Japanese.

Only 15 at the time, Mr. Myers did not understand about the political situation with Japan, and so when he heard over the radio that Earhart had been picked up, he was elated and thought she would be returned to the United States promptly. But to his surprise in the coming days and weeks, the media did not mention a word about Earhart's still being alive, much less her being captured by the Japanese.

On several occasions in the ensuing years, Mr. Myers said he received death threats over the telephone from unknown callers, who ordered him to "keep your mouth shut about Earhart!" The secret knowledge he had regarding the Earhart disappearance soon became quite an emotional burden for him, one he would bear until the early 1970's, when he saw the news footage of Irene Bolam.

A short time after seeing the news footage, Myers contacted Joe Klaas and obtained Irene Bolam's phone number. Afterward, he made the first of many calls to Bolam that would, according to him, span the next four years. In his book, Myers details some of his conversations with Bolam, which were very revealing. Although the one thing that Mrs. Bolam never did was to directly admit to Mr. Myers that she was actually Amelia Earhart.

But the fact that the calls, which consisted mostly of reminiscences of the times Myers and Earhart had at the Oakland Airport, went on for four years is extremely telling, if true. In other words, if Irene Bolam wasn't Amelia Earhart, why did she continue a four year old long telephone dialogue with Robert Myers?

Finally, in 1982, shortly before Mrs. Bolam died, Mr. Myers reported that she and Mr. Myers had a clandestine meeting in New Jersey. She was driven to the rendezvous place in a chauffeured car, and when the vehicle had stopped, she sent the driver away. She then waved to Mr. Myers to approach her car. Mr. Myers disclosed only part of the conversation in his book, but he did indicate that Ms. Bolam repeated her request to him to tell his story. She would not confirm that she was Amelia Earhart, however told him, "the things you say are all true".

Mr. Myers concluded his book with the remark: "In this last meeting, she again told me to tell my story, which is actually her story. Now I have done so".

As mentioned at the beginning of this chapter, Mr. Myers' account presents some serious conflicts: 1) According to Mr. Myers, the Japanese tried to force her to land on or near Hull Island. This is unlikely, as Hull Island was controlled by the British and would have been one of the first places Earhart would have wanted to land. 2) Mr. Myers reported that Earhart ditched several miles west of Sydney Island. Actually, there is strong evidence that Earhart ditched to the north near Mili Atoll in the Marshall Islands. 3) Myers maintained that Earhart had transmitted on her radio for nine days after her ditching. Radio experts employed by Lockheed Aircraft indicated almost immediately after Earhart's disappearance that the radio could not work if the craft ditched at sea, as the transmitter was located under one of the wings. 4) Mr. Myers reported lengthy radio transmissions over a nine day period, which no other radio operator has reported hearing.

As a result, most researchers tend to discount Mr. Myers' account. The devil of it is that in many other ways, Myers' account appears very plausible.

One possible explanation for points in Myers' account which seem outlandish or conflict with known facts, is that Myers wasn't telling untruths: he was only fifteen years old when he had his encounter with Earhart and didn't fully understand some things.

For instance, Myers mentioned that Fred Noonan was known to go by two names, Fred and Bill. He reported that the mechanics at the airport argued over which was his real name. It may well be that Myers and the mechanics didn't know that the two men were total ringers for each other and mistook them as one and same person. Or it may even be that the mechanics knew and pulled young Myers' leg as a joke.

Another misperception may be the incidents with GP. To a tender, impressionable kid of fifteen, the tall, aggressive, and frequently abrasive Putnam would definitely seem a bit sinister and intimidating.

Even the incident Myers described in which Putnam tried to run him over may actually have been just an mild accident in the failing light of the dusk. Not realizing that Putnam had inadvertently gone off the road due to the bad light, the spooked and understandably paranoid youngster may thought it was intentional.

Another small point is that some accounts inaccurately maintain that Myers reported AE had radioed to him directly. But Myers account as published in his book clearly indicates that he had stumbled on the broadcasts and merely monitored them, most of which were to the ITASCA.

All in all, Myers' account is a provocative mixture of the plausible and the unbelievable. His material seems far from fully reliable and one can't quite embrace or discard it. But his observations of day to day activity at the Oakland Airport are what seem most believable; it is his assertions re other matters, particularly radio transmissions that are open to huge eyebrow wiggling.

What is the truth regarding Robert Myers' account? Is it startling fact or the product of the imagination of a boy who used to visit the Oakland Airport?

Chapter 17

The British Connection

In 1987, Mr. James Donahue detailed the results of his investigation into the disappearance of Amelia Earhart, in his book, "The Earhart Disappearance: The British Connection". The centerpiece of his investigation was his discovery of a British involvement in the Earhart disappearance.

Mr. Donahue was a senior engineer for the Northrop Corporation, where he conducted system analyses of complex aviation technologies. A subspecialty of his was aviation history, for which he was a consultant to both Northrop and Lockheed Corporations.

At some point, he got involved in trying to find a Lockheed Model 10E aircraft as a display for a museum. This search stimulated Donahue's interest in the disappearance of Amelia Earhart, which culminated in his book.

The British involvement which Donahue identified, took several forms. One major form that Donahue discussed in the book was to closely work with Earhart in taking clandestine photos of the airport in Asmara, a city in the East African country of Eritrea. Eritrea was one of the areas annexed by Italy during their 1935-36 campaign in Africa, and Asmara was reputed to be the largest airfield in Africa. The airport at Asmara was "off limits" to the Earhart flight and was considered a strategic location by the British, who needed intelligence about this facility.

According to Donahue, there was evidence to believe that Earhart had landed at the RAF base in Khartoum, where a camera and film were installed in her plane in a compartment under the flight deck. Earhart then made an oblique pass near Asmara on her way East, took photographs of the large airport there and after the landing at Karachi, the camera and film were removed by RAF personnel.

Another element of the British involvement was a second flight over the Central Pacific by a British aircraft, which originated from Oakland, CA and, according to Donahue, flew from east to west. This flight apparently occurred at the same time as Earhart's Lae to Howland flight in early July 1937. Donahue asserted that the occupants of the craft did not fit the descriptions of Earhart and Noonan, and that the craft was made of wood, not metal, as was the Lockheed Electra

In addition, Mr. Donahue developed important evidence indicating that during Earhart's flight from Lae New Guinea to Howland Island, there was a British officer stationed on Hull Island, who was equipped with a powerful short-wave radio. The man Mr. Donahue identified was one John W. Jones, Captain, Royal Navy Reserve.

Mr. Donahue's thesis was that Captain Jones was assigned to support the British overflight of the Central Pacific, but also became involved in the Earhart flight after her disappearance. Additionally, there were ships of the New Zealand and British navies in the area, assigned there according to him to support the East to West British Flight.

Mr. Donahue hypothesized that Earhart had made a pass over the Marshall Islands and then headed for Howland. Before she reached Howland, according to Donahue, Earhart was ordered to bypass Howland and land "wheels up" on the shore of Hull Island, to the Southeast.

After Earhart ditched on Hull Island, Captain Jones, who was the resident manager of the island, retrieved Earhart and Noonan from the wreck of their plane. Afterward, Mr. Donahue maintained, Captain Jones spirited the two aviators to nearby Sydney Island and later to American Samoa, where Noonan died by misadventure and Earhart due to dysentery.

Donahue also speculated that Captain Jones was the originator of some of the SOS transmissions credited to Earhart after her disappearance. This may well be, as Pan American reportedly triangulated one of the messages as having come from the area of Hull Island.

Although some of Mr. Donahue's final conclusions are problematic, many of his findings are valid and important, and some of the material he developed without a doubt sheds light on the Earhart mystery.

A significant part of Mr. Donahue's investigation involved review and analysis of the logs of various ships, in the archives of various countries, including the United States, New Zealand and Australia. It also involved interviewing scores of witnesses. An impressive accomplishment, and an informative volume, Mr. Donahue's book is profusely illustrated with photos, maps and charts and is heavily documented.

David K. Bowman

Mr. Donahue also relied very much on the assistance of Paul Rafford, a 1930's radio expert and former Pan American Airways flight crew member, whose paper on Earhart in the Smithsonian Air and Space Museum caught Donahue's eye. In his computer-assisted study of Earhart's reported radio transmissions, he concluded that Amelia Earhart actually passed Howland Island and ditched in the area of Hull Island, which is south of Howland Island. In recent years, however, Mr. Rafford has indicated he no longer believes this to be the case, and is writing a book to outline his new findings.

Although not all of Mr. Donahue's conclusions are probably valid, a lot of his supporting evidence seems solid, and his discovery of the involvement of Captain Jones is a significant contribution to the solution of the Amelia Earhart mystery.

Chapter 18

Lost Star

One of the most unusual books that have come out about Amelia Earhart's disappearance was "Lost Star," by Randall Brink. It was the first and only book to date, to extensively discuss details of Earhart's secret mission for the government. Brink also presented witness testimony to buttress some of his points.

After carefully laying out the sociopolitical environment of the 1930's, Brink drew a brief but concise picture of Amelia Earhart's early career. He then carefully pieced together the arrangements for Amelia Earhart's last flight, using documents he obtained through the Freedom of Information Act and the testimony of several selected witnesses.

Brink used copies of White House and Cabinet memoranda to document how the U.S. Government decided to annex several of the Line Islands, of which Howland Island is one, in order build airstrips on them for defense purposes. He also introduced a subtly worded memo from FDR in which FDR tacitly ordered the U.S. Navy to involve itself in Earhart's flight.

One of Mr. Brink's important contributions to the record was his locating and interviewing Robert T. Elliott, Lloyd Royer, and Walter McMenamy.

A Lockheed employee in the 1930s, Mr. Elliott's account is discussed earlier in this book. His testimony, along with that of Lloyd Royer, raises the possibility that a switch occurred with Earhart's plane, and that the U.S. government was indeed deeply involved in the flight.

Walter McMenamy organized Earhart's private ham radio network and was an important member of her team. His testimony further proved government involvement:

"After she crashed in the Hawaiian Islands . . . the Navy and Coast Guard completely took over the flight. Amelia herself made no decisions anymore, and we had no contact with her. I used to see her at least a couple of times a week, but after the Luke Field episode, I never spoke with Amelia again. U.S. [FBI] agents called us to a meeting at a Los Angeles restaurant, and swore Karl [Pierson] and I to secrecy about the flight and the radio signals we received up on the [Beacon] hill".

Mr. Brink further located and interviewed Mr. Lloyd Royer, Earhart's unsuccessful suitor. During 1935-36 Mr. Royer had worked at Lockheed and knew quite a bit about preparations for Earhart's flight.

In an interview in 1977, shortly after the 40th anniversary of Amelia Earhart's disappearance, Mr. Royer said he had "worked on her first Electra, but the final flight was made in a different plane".

Another important witness that Mr. Brink located was Carroll F. Harris, a retired navyman who had, while on active duty, been assigned to microfilm highly sensitive files in the office of the Chief of Naval Operations and the Office of Naval Intelligence. For obvious reasons, he was unable to take any notes during his work, but he apparently had a good memory. He recalled that the Earhart files took up three-quarters of a file drawer and covered the following:

1. The navy's reasons for its clandestine participation in the Earhart world flight and speculation about the possible impact (including political repercussions) of disclosure of any details of the mission. One motive indicated in the mission was an overflight to photograph the Truk Atoll area in order to get intelligence on Japanese shipping activities.

2. Details of the precautions taken to keep Earhart's actual route from Lae to Howland secret. The files indicated she was to take a different, longer flight path that her publicly announced one, which was made possible by the Wasp Senior engines which had been installed in her ship.

3. Specifications for modifications to the Model 12A which was substituted for Earhart's original Model 10E after the Luke Field crash.[30]

---

[30] Yet another reference to a switch of aircraft. Did a switch occur?

4. Complete details, along with photographs, of the installation and operation of the Fairchild aerial survey cameras in the belly of the Lockheed Model 12A. [31] Also information on modifications to the ship's electrical system so that it could handle the increased load placed on it by the surveillance cameras.

5. Documentation on the installation of powerful, new Bendix radio equipment, which had greater power and range than anything commercially available at the time. Also information on a special high frequency direction finder that was installed on Earhart's craft, along with extra, more powerful batteries.

6. Details of arrangements for ONI (Office of Naval Intelligence) personnel to recover the film and equipment secretly and quickly upon completion of the mission.

7. Plans for part of the preparatory work on the Electra to be completed at Miami, with the remainder of the work to occur near the end of the flight before the departure from Lae.

8. Details concerning storage of the aircraft at the Alameda Naval Air Station. The file also included a list of work to be performed there. It indicated that some work was also completed at March Field and the Lockheed facility in Burbank.

Brink speculated that there had also been a presidential file on the Earhart disappearance maintained in FDR's Map Room, comprised of twelve boxes of material. There was indication that these boxes ended up in the custody of Henry Morgenthau, the Secretary of the Treasury. Brink indicated, although didn't cite his source(s), that the presidential file contained information on Earhart's fate and detailed records of cabinet meetings held in July 1937 to deal with Earhart's disappearance.

Brink concluded his section on these files by indicating that, to date, the Treasury Department had not granted an FOIA request for the twelve boxes of "Morgenthau Files".

Another interesting bit of new evidence adduced by Brink was an aerial photograph showing an airplane with only one wing and parked on a Japanese airfield on a small island in Maleolap Atoll. It's possible that the aircraft in this photo was Earhart's. The problem is the testimony of

---

[31] Ibid.

Thomas Devine. Mr. Devine's testimony credibly suggests that by 1944, at least, the Lockheed Electra was on Saipan in a building at Aslito Field. This is one more maddening contradiction that one encounters when trying to plumb the Earhart mystery. The probable final answer here is that while the plane bears a little resemblance to Earhart's plane, but it is not. According to Dr. Alex Mandel (see "Interception" for more about Alex), the plane is in all likelihood a damaged Japanese Mitsubishi bomber.

Mr. Brink also discussed the Morgenthau Transcript, as well as publishing a complete reproduction of that document (to be discussed in the next chapter). The only other book to publish a reproduction of that document so far was "My Courageous Sister," by Muriel Earhart Morrissey, which was also the first book to mention the Morgenthau transcript.

One other intriguing, if not startling, bit of evidence that Brink presented was a copy of a telegram received by George Palmer Putnam on August 28, 1945.

The telegram had been sent, very tellingly, via high priority government channels and the surviving copy was typed on a Speedletter form, with the text of the original message transcribed below. It was very telling that there was no name of the sender.

The message read:

"Following message received for you from Weihsien via American Embassy, Chungking:

"Camp liberated; all well. Volumes to tell. Love to mother (*).*
    Eldred D. Kuppinger
    Assistant Chief
    Special War Problems Division
(*) Signature omitted.

For some years, researchers and investigators have believed the telegram had been sent by Amelia Earhart. That certainly seemed to be a logical explanation. However, in 2001, TIGHAR (The International Group for Historic Aircraft Recovery) conducted an investigation to verify the genesis of the telegram.[32]

---

[32] Mike Campbell and Thomas Devine, "With Our Own Eyes: Eyewitnesses to the Final Days of Amelia Earhart," (Lancaster, OH: Lucky Press, 2002) pp. 129-130.

Surprisingly, they found that the communication had been written by one Ahmad Kamal, a Turkish writer and world traveler, who had been interned at Weihsien between 1942 and 1945. Ironically, Kamal, who was also a pilot and kept his plane in Burbank, California in the 1930s, had met Amelia Earhart and George Putnam during that period.

According to Kamal's son, Kamal had an agreement with GP to look after Kamal's mother while Kamal was in China. The telegram which GP had received in August 1945, then, was, according to Kamal's son, sent by Kamal to GP to let GP know that he had been liberated.

Furthermore, Kamal maintained that Earhart had never been held at Weihsien to his knowledge and he insisted that had she been there, he would have heard something about it.

The Weihsien telegram would now seem to be fully explained. However, what still raises questions in this writer's mind is why the U.S. government classified the telegram, if it was from a prosaic source, and why they kept it classified for 42 years.

The telegram was released during the week of July 2, 1987 when the media was running stories relating to the 50th anniversary of Amelia Earhart's disappearance. It is of special note that the telegram had been declassified by a State Department section identified as "Special War Problems Branch" and the file which had contained the telegram was labeled "Earhart, Amelia," with the file number "PW/8-2145". The "PW" designator was used only to identify prisoners of war. Why all of the special handling and special attention—for a telegram from a writer?

Aside from simple error, there is only one other possible reason for classifying the Weihsien telegram that comes to mind: whoever classified the telegram knew or suspected Amelia Earhart was still alive, and made the same mistake that researchers later made. They assumed the telegram was from Earhart.

Chapter 19

One Night on Saipan

One of the most provocative accounts to come out of WWII regarding Amelia Earhart's disappearance is that of Mr. Thomas Devine. His experiences are detailed in "Eyewitness: The Amelia Earhart Incident" and "With Our Own Eyes".`

Mr. Devine was a member of the 244th Army Postal Unit, which came ashore on Saipan after the island had been secured by the allies in July 1944. Comprised of 27 enlisted men and two officers, the unit was bivouacked in southern Saipan on Cape Obiam. From their camp, they had a clear view of attacks on nearby Tinian Island, as well as activity at Aslito Field, less than a half mile north.

Early on their first day on the island, Devine's commanding officer, Lieutenant Liebig, instructed Devine to drive him to Aslito Field, which had only recently been secured. The two started north, but were stopped by an MP (Military Police), who told them that they couldn't approach Aslito Field from the south. "We have orders to report there," LT Liebig told the MP. The MP took down the names and serial numbers of the two men and gave them directions so that they could approach the airfield from the correct direction.

As the two men neared the airfield, they encountered more MPs and wondered what was happening. Reaching the airfield, they drove around a hangar and saw a group of marine officers standing in front of the nearby administrative building. They parked along the side of the hangar and got out. In front of the hangar stood a group of enlisted men.

Devine and Liebig immediately noticed that the leader of the group was a man in civilian clothes, with an open throated white shirt. Liebig and Devine approached the hangar door, asking one of the marine guards where they should report.

Before the guard could answer, the man in civilian clothes stepped toward them, raised an arm and admonished them, "This is off-limits!"

LT Liebig told the man he had been ordered to report there and the man suggested that Liebig and Devine go to the nearby administrative building. The two walked over to that building. Devine remained outside, and a moment later, after Liebig was in the building and out of earshot, one of the nearby marine officers shouted, "What do you mean it's off-limits? We know Earhart's plane is in there! Our men laid their lives on the line, and now they won't even get credit for finding the plane".

Someone else said that a Wally Greene had identified the plane in the hangar as Earhart's. The hangar was now padlocked and guarded. "Our Major Wallace Greene?" asked an officer.

"Our *Colonel* Greene," was the retort from another officer. "He was just promoted". Another officer cautioned against "sounding off," but a moment later, the man in the white shirt walked toward the group of officers.

"I told you you'd get into trouble," the second officer muttered to the officer who had been shouting. The man in the white shirt asked the officer for his identification and returned to the enlisted guards in front of the hangar. Someone said that the place to settle the matter was inside the administrative building. The officers abruptly agreed and filed into the structure.

After listening to this strange exchange, Devine walked back around the side of the hangar to check on the jeep, and encountered another guard. The guard told him he'd have to move the jeep because the hangar was off limits.

"Is it true that Amelia Earhart's plane is inside?" Devine asked the guard.

"Yes, but for the love of God, don't say I said so!" the guard said. "I don't know why the hell they want to keep it a secret!" he muttered in exasperation.

A moment later, Liebig came out of the administrative building and Devine met him with the jeep. They left the airfield.

Later that same day, Devine was introduced to a marine who had been a guard at the hangar at Aslito Field. "They're bringing up Earhart's plane," the marine told Devine.

"Whose plane?" a fellow soldier named Fritzler asked in disbelief.

"Earhart's," the marine repeated.

"Devine did you hear what he said?" Fritzler asked.

"Yes, I heard," Devine replied and then asked the marine "What to you mean, "Bringing up the plane?"

The marine did not answer, changing the subject in obvious discomfort. A couple of hours later, Devine recounted, LT Liebig approached and reminded the men that no one was allowed to leave the area. Devine thought that was strange, as there were still a few scattered Japanese soldiers left hidden on the island and nobody was anxious to leave the security of their compounds.

Devine stated he soon learned that the reason for the order to stay put had nothing to do with the danger of hidden Japanese soldiers.

Almost immediately, he saw and heard two single engine military aircraft fly overhead and land at Aslito Field, after which there was the huge roar of engines as another craft apparently took off from the field. A few moments later, engines thundered overhead and Devine looked up. He saw what he described as "a large, twin-engine, double-fin civilian plane". Devine carefully noted the number on the craft: NR16020. The aircraft, which had apparently just taken off from Aslito Field, flew over Devine's compound and returned to the field.

That night after dusk, burning with curiosity, Devine decided to risk approaching the off limits area for another look. He and a friend, PFC Paul Anderson, slipped through the jungle to Aslito Field.

Devine reported that he and his friend spotted the plane that they had seen that afternoon, parked in front of a roofless hangar, in the southwest corner of the field. As Devine and Anderson approached, a photographer was taking photos of the aircraft. He was too far away for Devine to tell whether he was military or civilian. Abruptly, the photographer stood up and ran off.

This is the only part of Devine's account which seems a little unbelievable. Earlier in the day, authorities seemed quite anxious to keep all unauthorized personnel away from the airplane. According to Devine, the area had previously been closely guarded by MPs. But now, after dusk, Amelia Earhart's aircraft stood unguarded.

Near the ship, Devine and Anderson noted a group of cans containing some sort of fuel. They stealthily approached the plane and carefully looked it over, noting the following:

- The number on the wing, which was NR16020
- The edge of the wings had once been painted a color which was now yellowish orange
- The words "Hamilton Standard" were imprinted on the propeller
- The tail section was distinctive due to its two twin fins

- The left tire was flat, causing the ship to lean sharply to the left
- The aircraft was made of aluminum, which was now dull due to pitting

The inspection of the exterior done, they then decided to get into the aircraft to inspect the interior. Moments later, Devine and Anderson heard a click and looked up to see that the photographer was back and had just taken a photo of the both of them as they stood looking at the airplane.

Before either man could do or say a thing, the photographer again ran off. Simultaneously, two men came out of the hangar next door, where the MPs had been stationed earlier that day. One of them was dressed in a civilian white shirt and Devine immediately recognized him as the man he had seen that afternoon giving orders to guards at the hangar. Devine motioned his friend to follow him into the nearby darkness. They both crouched in wary silence and watched the figure of the photographer in the distance, waving the other two men back to the hangar they had just left.

"We'd better get the hell out of here," Devine whispered. They didn't waste any time slipping back into the jungle.

A short time later back in camp, after a hot bath, Devine was just returning to his tent, when he heard an explosion and could see a fire in the distance near Aslito Field. Sprinting as fast as he could, he reached a spot near the hangar and couldn't believe what he was seeing: NR16020 was in flames.

Devine returned to his camp and went to sleep for the night, making it a point to write down the number he had seen on the wing of the twin engine plane. Two days later, he recounted, he awoke and discovered his pants were missing. They had contained his wallet, where he had placed the slip of paper containing the airplane number.

Fortunately, Devine hadn't forgotten the number—the strange events he had witnessed at Aslito Field would prevent that. He wrote it down again later.

Devine ended his description of this incident by stating that he had concluded that he and his friend had happened on an intelligence-related operation. It was clear to him that the guards had been removed so that nobody would witness the final preparations for the destruction of Amelia Earhart's plane. He also speculated that the original plan had called for flying the plane out to sea and ditching it. But due to the flat left tire and possibly his and Anderson's unexpected presence at Aslito Field, Devine reasoned that the plane was immediately burned to destroy its identity.

Devine reported that the wreckage of the Electra remained at Aslito Field for the rest of his time on the island. Pieces of the ship's aluminum fuselage, according to him, were removed from time to time by GI's who were newly stationed on the island[33].

If Devine is correct, then the many attempts that have been and continue to be made to locate Earhart's famous airship in various places in the Central Pacific are ironically pointless. According to Devine, sometime after Saipan was secured, construction eventually started on a large airport on the site of Aslito Field. During construction, any airplane wrecks that couldn't be removed were buried in large holes dug by bulldozers. Without a doubt, the remains of Earhart's Electra rest today somewhere under the foundation of the Saipan International Airport.

Intriguingly, Devine disclosed that he recognized the man in the white shirt, who had placed the hangar at Aslito Field under security and later ordered the burning of NR16020, as newly appointed Secretary of the Navy James Forrestal.

Near the end of Devine's tour of duty on Saipan, in July of the next year, Devine had a second provocative experience. He and a friend, PFC John Boggs, went on a long, exploratory walk through southern Garapan, during some off time. Devine carefully described the landmarks he remembered as they progressed through their outing. At length, they came upon an Okinawan woman hanging out laundry near a sheet metal building. Fortunately, a Japanese-American serviceman happened, apparently from the temporary camp at Susupe, and he ended up translating for Devine and Boggs.

A conversation was soon started with the woman, who according to Devine, was dressed in a white cotton hospital type dress and appeared "timeless".

The woman first asked Devine about the insignia on his sleeve, and the interpreter told her Devine was a sergeant in the U.S. Army. Abruptly, the woman then said through the interpreter that she knew where two white people were buried, one of whom was a woman.

Devine discounted this at first, as he knew that the islands had been off limits to outsiders for years before the war. The woman then said that she and her husband, who was fishing nearby at that moment, had lived in the area before the war.

At that point, the interpreter told Devine that he had been sent from Camp Susupe to retrieve the woman and a companion. Devine then asked the woman how the two white people came to Saipan.

---

[33] Thomas Devine with Richard D. Daley, "Eyewitness: The Amelia Earhart Incident" (Frederick CO: Erennaissance House, 1987) pg. 56-57.

LEGERDEMAIN

The woman pointed upward and said, through the interpreter, they had "come from the sky". She then grabbed Devine's arm and led him to a spot about 100 yards away, not far from a small temporary cemetery. At that moment, another woman, her companion, appeared at the door of the sheet metal building.

Gesturing toward the ground, the woman fell to her knees and said through the interpreter, "They are buried here, beneath us!"

"Who buried them?" Devine asked.

The woman became very agitated and finally began screaming and pointing at the interpreter. The interpreter tried to talk to her, but she continued to scream at him and began to strike him.

"Why is she hitting you?" Devine asked the interpreter in surprise and concern.

"She says the Japanese killed them. She says I'm just as guilty as the rest of them". The woman hadn't realized that the interpreter was a Japanese-American and guiltless . . .

Finally, the interpreter was ready to take the two women back to Susupe. The Okinawan woman pleaded with Devine to take her with him. Evidently, she did not want to return to Susupe. Feeling really bad about the situation, Devine had to regretfully inform the woman that he had no authority to help her.

Shortly after that incident, Devine heard that the area where the Okinawan woman had taken him near the temporary cemetery was being excavated by MPs. At the time he could only assume they were routinely looking for the remains of American soldiers. A couple of weeks later, the war was over and Devine was sent back to the United States. But he never forgot his experiences on Saipan, both at Aslito Field and in southern Garapan near the temporary cemetery.

On August 12, 1945, word reached Saipan that the war was over. With every ship and boat moored near the island blowing their whistles and horns in celebration, Thomas Devine received orders to report to the replacement depot to be shipped home for discharge. He was extremely excited and happy.

Shortly after Devine was assigned a cot at the depot, a man in navy-type of dungarees and a t-shirt appeared and instructed Devine to go with him.

"Where are we going?" Devine asked.

"You're going by air," the man said. Devine was elated at this and grabbed for one of his duffel bags. At first, the man in dungarees wouldn't grab the other bag, but finally assented, when Devine said he wouldn't go without his two bags.

Outside the building, the man in dungarees commandeered a truck, showing the driver his credentials. Devine thought little of this at the time and got into the truck. As the truck drove towards Tanapag Harbor, Devine asked his escort why he was going by air.

"You're going to Honolulu," the man said.

Devine then told the man several personnel-related concerns he had, including delays in Honolulu.

"It's all taken care of," Devine's escort assured him.

Devine argued for a few moments more, before the man finally said to him, "All right, I'll level with you, but you gotta get on the plane."

"Okay," Devine prompted, "let's hear it."

"You're going to Honolulu for a briefing."

Devine then balked in incredulity and demanded to be taken back to the depot.

"You can't go back, Devine!" the navy man said, just as the truck reached a pier in the harbor. "They're waiting for you!" He suddenly blurted out, "You know about Amelia Earhart!"

Nearby, a PB2Y seaplane was tied up, an officer standing on the pier in front of it. The officer, possibly the pilot, was extremely brusque, ordering Devine to leave his bags at the pier.

Devine told the officer he couldn't leave without his orders and the man snapped back at him, "I'll get your orders, but get on that plane!"

Just then, Devine noticed that the navy man was running up a nearby hill.

"To hell with this!" Devine said, grabbed his two duffel bags and marched back to the road where he managed to get a ride back to the depot. He was later shipped out that day, but remained apprehensive, as his name was constantly called out over the PA system.

It gave Devine pause to think when, a few years after WWII, in 1949, James Forrestal died under very unusual circumstances at Bethesda Naval Hospital. Forrestal fell to his death from a sixteenth floor hospital window, and although his death was ruled a suicide, it nevertheless remains questionable to this day. This incident caused Devine avoid any inquiries into his Saipan experiences for years.

Years later, in 1960, Devine finally initiated an investigation into Earhart's disappearance. One of his main concerns, initially, was to locate dental records for Earhart and Noonan, so that any remains that might be found during a future excavation of the gravesite shown him by the Okinawan woman in 1945 could be properly identified. Over the years, he had come to be convinced that the gravesite was that of Amelia Earhart.

Devine searched various places in the Boston area relentlessly, but with no success.

On October 3, 1960, Devine received an unexpected call from the Office of Naval Intelligence (ONI) requesting an interview with him. Readily agreeing to this, Devine had a lengthy session with ONI Agent Thomas Blake. He related everything that he had experienced or uncovered and placed his collection of photos at their disposal. Oddly, Devine wouldn't hear the results of the ONI's subsequent investigation of his revelations until January of 1963.

After the interview, Devine resumed his inquiries in the Earhart disappearance. In May of 1961, he applied for permission to visit Saipan, and the government turned him down. At about the same time or shortly thereafter, Fred Goerner applied for permission to go to Saipan and was approved. When Mr. Devine learned of this not long after, he was not pleased. He contacted the Hartford ONI Office and requested permission to accompany Goerner at his own expense. That permission never came.

In the summer of 1961, he went to visit Muriel Morrissey, Earhart's sister, in West Bedford, MA.

Arriving at the Boston train depot, Devine encountered a man who told all the surrounding travelers that he had a cab that he was taking to West Bedford and that someone could ride with him for free. Many travelers tried to accept the offer, but the man kept pressing Devine to accept. Devine was reluctant, but found himself a short time later, in the cab—by himself. The cabby would allow no other travelers to ride in the cab. Devine said he was never asked his destination and little conversation passed between him and the cabby. A while later, Devine was deposited in West Bedford a block from Mrs. Morrissey's home.

With little prompting, the cabdriver gave Devine directions to Mrs. Morrissey's home.

Mrs. Morrissey greeted Devine cordially, telling him that she had notified the Navy of Devine's visit, and this was understandably puzzling to the ex-marine. He wondered why she would want or need to do that.

Although Mrs. Morrissey was unable to give Devine any help in securing a copy of her sister's dental records, the visit was lengthy and enjoyable. At the end of the stay, Mrs. Morrissey went to a livingroom window over which the shade had been drawn, and raised then lowered the shade. She said something about protecting the room from the effects of the sun. After Mrs. Morrissey excused herself to check on her mother, who live with her, Devine stepped to the window and peered out. He could see two men standing a short distance from the house, one of whom he recognized as the cabby!

Back at the train depot, Devine sought a meal in the depot restaurant. He noticed two men and a woman enter. The woman walked immediately behind the counter, where Devine was seated, and he looked on as she had a whispered word with the waiter and he handed her an apron. She put it on and began to wait on a nearby table. Almost immediately, Devine noticed the two men who had entered the restaurant with the woman. They were the two men he had seen outside the Morrissey home. Devine pretended not to recognize them. Then, strangely the two men tried to start what was, to Devine, a phony argument. The woman in the apron pressed Devine to stop the loud obscene argument, but he remained in his seat. Abruptly the two men made noises as though they were going to fight outside of the restaurant. Again, the woman in the apron urged Devine to intervene. Again, he sensibly remained seated.

A short while later, Devine stepped on the train, but noticed the cabdriver entered behind him. Quite shaken, Devine walked to the back of the car and exited it just as it started to move.

It was quite clear to Devine that he was being followed, but he simply wasn't sure why. Two years later, he got confirmation of this on a subsequent visit to the Hartford ONI office. He had dropped by to review ONI files on Earhart and was surprised to almost immediately recognize one of the people he had encountered that day as the woman in the apron in the Boston train depot.

As mentioned earlier, in January 1963, Devine was shown a copy of the report of the ONI investigation of his experiences on Saipan. He was shocked that the document was hugely biased, unfairly negative, inaccurate and dismissive. Interestingly, the report was classified Secret by ONI at the time. That in itself makes two statements.

First it infers that there was indeed at least something to Devine's experiences in 1944 at Aslito Field, as the Secretary of the Navy had been involved. Second, given all the biases and errors that Devine indicated he saw in the report, one can only think that the document was classified because it couldn't have withstood open scrutiny . . .

Mr. Devine tried again on subsequent occasions to get permission to go to Saipan, but was refused. He was quite frustrated when, finally in late 1963, Fred Goerner offered Devine the chance to accompany him on his next visit to Saipan at Goerner's expense. Devine jumped at it. The trip turned out to be frustrating and not overly productive for Devine, who later discovered that Ross Game of the Napa Register had footed the bill.

Subsequent to publication of "Eyewitness," Devine collaborated with Mike Campbell to write another book, entitled, 'With Our Own Eyes," which detailed his efforts at further investigation into Earhart's disappearance. One notable discovery was the explanation for the telegram from Weihsien POW Camp (see "Lost Star"). One thing that Devine did was to gain access to Forrestal's official calendar through the Freedom of Information Act. In Forrestal's diary entries, Devine found an intriguing record of a conference on July 19, 1944 at Forrestal's office between Forrestal and GP Putnam. That was just days after Forrestal's return from the Far East, and the inference was clear: Forrestal must have called the meeting to advise Putnam that his wife's plane had been found.

Another technique Devine used to investigate the Earhart disappearance, was to publish ads asking former military personnel stationed on Saipan in WWII with information regarding Earhart to contact him. One such ad was placed in Leatherneck Magazine in November 1993 and February 1994. As a result, a man only identified in Devine's book as "David F". sent a document in longhand on April 5, 1994 to an associate of Mr. Devine's, Henry Duda.

In his account, David F. claimed to have been a congressional page in 1944. David indicated in his account that pages were politically appointed and had to come from the families of congress members or other high officials. He stated that one day he had occasion to go to the Oval Office with a message, arriving during a conference between FDR, Vice President Henry Wallace, and Harry Truman. Entering the room, David heard FDR say, "They found the bitch's airplane," referring to Amelia Earhart. They then talked about it for a moment, after which Roosevelt suddenly said, "Destroy the airplane".

If true, this is an incredible revelation, it indicates Forrestal was acting upon the order of the president when he ordered the destruction of the Lockheed Electra that night long ago on Saipan. Moreover, FDR's caustic remark about Earhart is surprising. The two were supposed to be friends. However it could be that the Earhart affair represented an troubling problem for FDR if disclosed at the wrong moment. Only further data can help shed light on FDR's remark.

Mr. Devine's book also showcases the accounts of some two dozen WWII veterans who served on Saipan, and presents multiple corroboration of Forrestal's presence on the island and his involvement in the destruction of Earhart's plane.

In the end, Mr. Devine seems an unimpeachable witness and his account has the ring of truth. It will definitely figure in the denouement of this book.

Just before this book was completed, I had occasion to meet with noted Amelia Earhart researcher Ron Bright, and during our discussions, the subject of the Devine account was brought up. Ron indicated to me that he had thoroughly researched James Forrestal's schedule for 1944 and found no indication Forrestal was on Saipan in July 1944. In fact, he told me that he had found documentation that in early July 1944, Forrestal was at Newport, Rhode Island.

Ron was firm on the matter of Forrestal's location, which gave me pause to think. Was the man Thomas Devine had seen long ago on Saipan really Forrestal, on a totally clandestine visit, which had been left out of all official calendars? Or had Thomas Devine misidentified the mysterious man in a white shirt, who seemed to hold authority over the entire island?

In any event, there is solid evidence that Earhart's Electra was there on Saipan in July 1944, that there was somebody there who placed Aslito Field under a security cordon, and that somebody destroyed the Electra under orders from Washington. But who?

Chapter 20

An Unclaimed Letter in Jaluit

Another mystery regarding the Earhart disappearance surfaced in a small article in the Pacific Island Monthly, a magazine published in Sydney, Australia. Entitled, "POSTAL MYSTERY, UNCLAIMED LETTER FOR AMELIA EARHART," it was written by Carl Heine, a special correspondent to that publication and German missionary working in the Marshall Islands. The article appeared in the May 25, 1938 issue and was datelined Jaluit Atoll, March 17, 1938.[34]

The article ran:

"Here is a curious thing, on November 27, 1937, in the Jaluit Post Office, in the Marshall Islands (Japanese), among the unclaimed mail was a certain letter that attracted my attention. In its upper left hand corner was printed Hollywood Roosevelt Hotel, Hollywood, California . A little lower down appeared the postal date stamp with Los Angeles, California, October 7, 10PM, within the circle. Lower down in the usual place appeared the following startling address:

"Miss Amelia Earhart (Putnam)
Marshall Islands (Japanese)
Radak Group, Maleolap Island (10)
South Pacific Ocean

---

[34] Pacific Island Monthly, May 25, 1938.

"Written diagonally across one corner was this: 'Deliver Promptly.' On the back of the envelope, 'Incognito' was pencilled in very small, fine handwriting. The letter was unopened and consequently I have no idea of its contents. Now, it seems to me that anyone in the U.S. writing as late as October, ought to be well aware that Amelia Earhart had been given up for lost long before. Hence it would appear that the letter may have been written by someone desirous of hoaxing the public. Still it is possible that such may not be the case at all.

"Certainly the writer of the address on the envelope while making some errors such as anyone at a distance might make, displays a little more geographical knowledge of these parts than one would expect of anyone about to traverse the Pacific and would be passing this group at a distance of a few hundred miles.

"It is conceivable that Amelia Earhart may have told some trusted friend in America, before setting out on her ill-fated journey that she intended to take a look-see at the Marshalls, enroute, or that she might possibly do so if in any danger as she passed by. And it is possible that this hypothetical friend in Hollywood might think that Amelia had reached this group and might be lying low for some reason or other at Maleolap. It seems curious that anyone without specific interest in this group should know the name of that particular atoll, which is of no great importance. What the number (10) might mean in connection with that island, I have no idea".

Rollin Reineck noted in his book that one of Earhart's personal secretaries was living at the Hollywood Roosevelt Hotel in September and October 1937. He also noted that other researchers strongly suspect that there never was an unclaimed letter, but that the article was Mr. Heine's attempt, in code, to inform the world that Earhart was in the Marshall Islands, and on Jaluit, and that her plane was on Maleolap Atoll [Maleolap Island (10)]. The 10 in parentheses, it was suggested, referred to the model number of Earhart's plane.

Some five years after the publication of this amazing and cryptic article, early in WWII, Mr. Heine and his wife were executed by the Japanese. One wonders whether the two events were at all connected.

A further development occurred in the mid 1980s, when Buddy Brennan interviewed Mr. Heine's grandson, John. Brennan's book spelled the name "Heinie," which may or may not be an error. Mr. Heine remembered the article that his grandfather had written about the letter, but maintained that it was a postcard. He also said that the document was sent *by* Earhart and not *to* Earhart. Heine told Brennan that there was a reproduction of the document in his grandfather's article. Greatly excited by this revelation, Brennan asked Heine if he still possessed the article.

Unfortunately, according to Heine, he had given the article to a cousin years before and did not know if his cousin still had it. Brennan was unable to locate Heine's cousin during the rest of his stay on Jaluit.

Additionally, Earhart's secretary never claimed to have written the letter, and thus the article about the letter could be explained as an attempt in code by Heine to let the outside world know that the aviatrix had or was being held on Jaluit.

However, there is one haunting fact that begs explanation. That is, how would a missionary on an isolated South Pacific Island have known the name and address of the personal secretary of a world-renowned aviatrix? There are only a couple of possible answers to this question.

One answer is that the article about the letter was an encoded message and that Carl Heine wrote his article based on information he had secured locally. But from whom? Heine could only have gotten this information from one or two sources: 1) From Earhart herself on Jaluit, or 2) Some visitor to the island who had this information.

As to whom this visitor might have been, we can only conjecture. One name which is sometimes brought up by researchers is that of Vincent Astor (See chapter on the Nourmahal, "Secret Cruises.") Astor, though, has been reported to have been refused entry to the Marshalls by the Japanese during his spring 1938 cruise to the Central Pacific. Another thing is that Heine was known to be an interpreter for the Japanese governor of the Marshall Islands. In that capacity, he may have learned something about Earhart's secretary from some visitor who he was asked by the Japanese to interrogate.

The other answer is that the letter to Amelia Earhart in the Jaluit Post Office really existed, and Heine's article was, again, an attempt to tell the world that Earhart was or had been on Jaluit. Did GP have some inside knowledge of Earhart's whereabouts and have his wife's personal secretary write and send a letter to Jaluit, in the desperate hope of contacting the detained aviatrix?

Chapter 21

Notes in a Bottle

Another strange side-mystery in the disappearance of Amelia Earhart started the day before Halloween 1938 near Soulac-Sur-Mer, France. On October 30, 1938, a Mrs. Genevieve Barrat, aged 37, was walking on the beach near town and noticed a small bottle floating near the shore. Retrieving the bottle, Mme Barrat noticed that its capacity was about eight ounces, that it was closed with a cork and sealed with wax.

In the bottle she found a lock of hair and three documents. The first document contained only the words, "God guide this bottle. I confide my life and that of my companions to it" in French. The second document, also in French, was considerably longer and covered both sides of the paper. The third document was in shorthand.

Mme Barrat immediately took the bottle and its contents to the local police (Gendarmes), who, after reviewing the materials sent them on through channels, until they finally arrived at the National Gendarmerie headquarters in Paris (See copy of National Gendarmerie report in Appendix IV.) The French authorities eventually notified the U.S. State Department, which, ultimately, took no significant action. Their copies of the material ended up in the U.S. National Archives.

Below follows the complete text of the notes:

"Have been prisoner at Jaluit (Marshall) of Japanese in a prison at Jaluit. Have seen Amelia Earhart (aviatrix) and in another prison her mechanic (man), as well as other prisoners; held for so-called espionage of gigantic fortifications which are built at Atoll.

"Earhart and her companion were picked up by a Japanese seaplane and will be held as hostages, say the Japanese. I was a prisoner because I debarked at Mili Atoll. My yacht 'Viveo' sunk, crew massacred (3 Maoris), the boat (26 T) was supplied with wireless.

On reverse of paper:

"Having remained a long time at Jaluit as prisoner, I was enrolled by force as a bunker-hand on board 'Nippon Nom?' going to for Europe. Shall escape as soon as the ship is near the coast. Take this message immediately to the Gendarmerie in order that we may be saved.

This message was probably thrown off Santander, and will surely arrive at the Vendee towards September or at the least October 1938, remainder in the bottle tied to this one, Message No. 6."

A third document, in shorthand, read:

"In order to have more chance of freeing Miss Amelia Earhart and her companion, as well as the other prisoners, it would be preferable that policemen should arrive incognito at Jaluit. I shall be with JO . . . eux and if I succeed in escaping . . . for if the Japanese are asked to free the prisoners, they will say that they have no prisoners at Jaluit. It will therefore be necessary to be craft in order to save the prisoners of Jaluit. At the risk of my life, I shall send further messages.

"This bottle serves as a float for a second bottle containing the story of my life and . . . empty, and a few objects having belonged to Amelia Earhart. These documents prove the truth of the story in ordinary writing and shorthand and that I have approached Amelia Earhart . . . believed to be dead.

"The second bottle doesn't matter.

"I am writing on my knees for I have only a little paper, for fingerprints taken by the police. Another with thumb.

"Message written on the cargo board, No. 6".

The second bottle was never found, and the anonymous writer of the notes was never identified.

South African investigator Oliver Knaggs later carried out a thorough investigation of the discovered the existence of the notes after he discovered copies of them in a microfilm roll he had obtained from the U.S. National Archives.

The notes were sandwiched in among numerous bizarre notes and communications received by the U.S. government in the 1930s.

Subsequently, he mounted a lengthy trip to France, the United States and Guam to investigate the disappearance of Amelia Earhart.

Oddly, just before Knaggs and his wife left on their trip, Knaggs' briefcase was stolen. It contained Knagg's correspondence file which he and his wife and carefully assembled for the trip, as well as all of the contact information and notes they would need for the trip, along with the microfilm Knaggs had gotten from the U.S. National Archives. Was this just a chance theft—or did someone want to impede Knaggs investigation?

Undaunted, Knaggs and his wife proceeded with their trip as planned.

Knaggs later detailed his investigation his now difficult to find book, "Amelia Earhart: Her Last Flight". He was able during his trip to view all the materials in the French archives in Paris, as well as in the U.S. National Archives.

After he had reviewed the evidence, Knaggs carefully thought things over and made a list of "pros" and "cons" regarding the authenticity of the materials in the bottle.

Favoring the document's authenticity, for Knaggs, was a number of factors. Knaggs first cited the evident knowledge of South Pacific geography and of Japanese fortifications on some of them. Second, Knaggs felt the lock of hair found in the bottle may have been authentic, for although the report written in Paris in January 1939 at the Brigade level cited the color of the hair as "chestnut," the Gendarmerie Report described the hair as "light brown". Knaggs also found the writer's remarks describing Earhart as an aviatrix and Noonan as "her mechanic" to be so seemingly uninformed and unaffected that they almost had to be a mark of credibility. Also, the writer's reference to having been arrested because he had simply disembarked at Mili seemed credible in light of later hindsight about the Japanese fortification of that area. "How on earth could anyone have made such a statement unless he had been there?" Knaggs asks in his book.

Knaggs mentioned in his book that he had located a Monsieur Eric De Bisschop, a former French naval officer, who had, during a trip through the South Pacific in 1938 been briefly detained by the Japanese. According to Bisshop, his initial reception had been quite cordial until he mentioned that he had passed close to Mili Atoll. Abruptly, all friendliness evaporated and he was placed under arrest. He was crisply interrogated for several hours, but finally released after the Japanese could find no evidence of espionage activities.

Against the authenticity of the materials in the bottle, Knaggs noted several things. First of all was the fact that the writer did not give his name, although Knaggs conceded that the writer may have feared the bottle would fall into the wrong hands. The next factor arguing against the authenticity of the materials was the method of delivery: a sealed bottle washed up on a beach. This was definitely a rather melodramatic method of receiving such a disclosure. Finally, Knaggs listed as a negative factor, the lock of hair found in the bottle. He admitted that he had listed it as a "con" because he realized that, since he had not been able to see the actual lock of hair, its color could be argued either way.

In the end, Mr. Knaggs was fairly well convinced of the authenticity of the messages in the bottle. But frustratingly, he had been unable to locate the actual bottle or the lock of hair for examination.

Again, we are left with another strange tale with no final resolution. It is a tale filled with enough corroboratable information to be tantalizing but not conclusive.

Is there any truth to the strange notes and lock of hair once found in bottle on a seashore in France?

Chapter 22

The Ghosts of Garapan Prison

Yet another noteworthy attempt to fathom the mystery of Amelia Earhart's disappearance had its genesis in 1981 on the island of Majuro in the Marshall Islands. The investigation was detailed in a book entitled "Witness to the Execution," by T.C. "Buddy" Brennan, an excellent read.

Texas businessman Brennan had come to the Marshall Islands looking for WWII vintage Japanese warplanes that could be salvaged for display in museums. During his visit, in which the island of Majuro was his headquarters, Brennan made forays to various places in the vicinity, including Mili Atoll. To his frustration, all of the aircraft he encountered were badly deteriorated, and thus not usable.

But one thing salvaged the trip for him, a casual conversation he had with a local named Tamaki Myazoe, regarding the disappearance of Amelia Earhart. This captured the businessman's imagination, and when he wasn't hacking his way through deep jungle with a machete, looking for ancient Japanese airplanes, he was thinking about the disappearance of the famed aviatrix.

The day before he was scheduled to leave the Marshalls to go home, the Texas businessman wrote off his airplane search, at least on that trip, as a failure. But he had become interested in the Earhart disappearance and resolved to spend his last day on the island interviewing locals to find out what they knew of the mystery. Brennan would later mount four more expeditions to the Pacific in an attempt to solve the Earhart mystery.

Initially, Brennan was more interested in just locating Earhart's plane for retrieval and display in a museum. But as he studied more about Earhart herself, he became captivated. He had to find out what had happened to her.

On his next visit to Majuro, Brennan spent the entire time interviewing witnesses. The first witness he interviewed, and one of the most interesting, was John Heinie[35], grandson of Carl Heine, the missionary who published the article in the Pacific Islands Monthly about the unclaimed letter in Jaluit (discussed in the previous chapter). The grandson talked of the incident of the letter in the Jaluit post office, but, strangely, in his account, the letter was *from* Earhart and not *to* Earhart. Evidently Brennan wasn't aware of the incident, as he did not question John Heine about the descrepancy.

Brennan also interviewed other witnesses, whose consensus was that the Japanese had taken Earhart from Jaluit to Kwajelein, Truk and finally Saipan. With the visit at an end, the agenda for the next expedition was clear for Brennan: Saipan.

Early in the trip to Saipan, Brennan was taken to Garapan Prison, an old Japanese prison, on the edge of town. For many years, residents had told interviewers that Earhart and Noonan had been held in that facility by the Japanese. The Texas businessman wanted to see the place for himself. Once he had arrived there, he wished he hadn't.

By then, the old facility was in ruins and the jungle was beginning to reclaim it. Brennan's guide, Manuel Muna, also known as "Manny," showed him around. The atmosphere of the abandoned prison was brooding and eerie; about the place there hung an aura of misery and suffering. Brennan did not like the place at all; he could almost feel the unhappiness of the former prisoners.

It was as though their ghosts haunted the place. Amazingly, Manny, seemed cheerfully oblivious to such things, but Brennan couldn't shake a feeling of depression and unease. He wanted to leave, but had to stay as part of his investigation. He forged on, telling his readers, "This place gave me the creeps".

The first place they visited was a smaller building not far from the main one. It contained only four cells and was considered by the Japanese, according to Manny, to be their maximum-security building. Manny told Brennan that one Jesus Salas, a local farmer, was incarcerated there during the Japanese occupation.

According to Manny, when the invasion of Saipan started in 1944, Salas managed to escape from his cell during the shelling and confusion. Manny showed Brennan the cell which had been Salas', as well as the cells which Earhart and Noonan had reportedly occupied.

Brennan was to receive another surprise: Manny told him that Salas had reported seeing Earhart in the prison up until his escape in 1944!

---

[35] Brennan's book spelled the name, "Heinie," which may have been an error, since it appeared as "Heine" in the Pacific Islands Monthly.

According to Manny, his own sister had seen Earhart in the facility as late as 1941, while she was working in the area. After the war started, however, according to Manny's sister, the Japanese would not allow locals anywhere near the prison compound.

This information is very interesting if true, as by all other accounts, Earhart was either executed, died or deported by early 1938. For Earhart to have been on Saipan as late as 1941 or 1944, added a new dimension to the mystery.

When finally it was time to leave the eerie, deserted prison, Brennan was not at all sorry about it. He told himself mentally he would never come back. And he didn't.

After leaving Garapan Prison, they immediately went to the nearby home of Mrs. Nievas Cabrera Blas, who had an interesting story to tell. She related how, after the invasion of Saipan had started, the Japanese had ordered all the residents of the island to go to a series of caves that were on the islands, for shelter from the shelling.

However, for some reason she did not explain, Mrs. Blas remained in the area, and was working not far from a secluded site at which she said she witnessed Earhart's execution. She recounted that the Japanese had taken Earhart out of the prison, handcuffed and blindfolded, in a motorcycle with a sidecar to a spot not far from where she was working. There was a fresh hold dug in the ground. According to her, they removed the blindfold and tossed it into the hole. They then shot her in the chest and she fell into the hole. It was then that Mrs. Blas ran from the area, to avoid being seen by the Japanese. Later, she returned to the spot and found the hole filled in.

Brennan asked Mrs. Blas if she could lead him to that place. She said, "Oh, yes, it was underneath the biggest breadfruit tree on the island. I went there often to get the breadfruit".

A short while later, the small party was in a van driving toward the site. Eventually, it turned out that the breadfruit tree had been cut down and a parking log for heavy equipment built over it. Brennan asked Mrs. Blas if she was sure this was the site of the execution she had witnessed. She maintained that when she saw men cutting the tree down, she had mentally marked the site with other landmarks nearby.

Brennan, et al, decided it would take time to get permission to dig in the compound, and in the meantime, they decided to videotape Mrs. Blas' account for the record. At the wise suggestion of one of the team members, they adjourned to a nearby park where they could talk to Mrs. Blas discreetly.

Brennan ended up making a further visit to Saipan in order to excavate the site he had located through Mrs. Blas. Unfortunately, after he returned to

Saipan, he found out that he could only excavate for one day a week—Sunday. It turned out that the area in the parking lot that he wished to excavate was a high traffic area and the authorities there adamantly insisted that Brennan could only excavate on the one day of the week when there was no traffic.

Having no choice, Brennan organized an intense, one day dig. Near the end of the dig in the late afternoon, a piece of fabric was discovered in the bottom of the hole. It was immediately shown to Mrs. Blas, who was nearby. With a show of great emotion, she said through her granddaughter Rosa, her interpreter, that it looked like the blindfold she had seen the Japanese remove from Earhart before executing her over 40 years before.

Brennan and his fellow team members were stunned. They felt they had successfully found a historical artifact. But one large problem still remained with the excavation: no human remains had been found buried with the blindfold.

At this point, Brennan's time on Saipan ran out and he reluctantly returned to the United States. He concluded that Earhart had ditched near Mili Atoll and had been picked up by the Japanese. From there she was taken to Saipan, where she was incarcerated for seven years before being shot at the beginning of the 1944 invasion. Since he had not found remains buried with the blindfold, Brennan concluded that Earhart had been buried somewhere else, but could not say why.

The Brennan account introduces several new elements into the Earhart mystery: Earhart's possible presence on Saipan as late as 1944, the only first hand eyewitness account to date of an actual execution, and an enigmatic blindfold.

On the face of it, the blindfold doesn't corroborate the account of the execution—not completely. It only corroborates that Nieva Cabrera Blas may have seen something that *appeared to be an execution.* It would seem that long ago at the beginning of the invasion of Saipan, a grim scenario might have been enacted by the Japanese in a secluded place on the island, with only one person as an audience. What had Mrs. Blas really seen? An execution…or a charade?

Other evidence could answer this question. During their visits to Saipan in the late 1960s, the Kotheras encountered another witness to what may have been the same incident. The witness' name was Anna Magofna, a resident of Saipan, who claimed to have been on the island in the summer of 1937.

Amazingly, she claimed to have witnessed an execution near the same breadfruit tree described by Nieva Cabrera Blas in her account. But Magofna reported that she *two* white people executed by samurai sword, claiming that she was the only witness.

David K. Bowman

Could these two accounts have concerned the same execution? If so, why did each woman think they were the only witness and why the other discrepancies in the two accounts?

Moreover, another element enters the picture which may explain the absence of a body in the grave which Brennan excavated. Later in this book, in Chapter 24, the case of Everett Henson and Billy Burks is discussed. They were two marines who were ordered to secretly excavate a grave *near a breadfruit tree in a cemetery not far from Garapan.* They reported finding remains which they turned over to a marine officer (see Chapter 24 for complete details). Could this explain the absence of a body in Buddy Brennan's excavation?

If that isn't enough, there are at least a couple of other accounts of different executions occurring near the large landmark breadfruit tree. Logically, they couldn't all have been the same execution. Either that breadfruit tree was just a damn busy place, or there is more than a little confusion afoot here.

While Buddy Brennan's experience was one of eerie impressions, another researcher, Eugene Sims, had one that may have been literally paranormal. In the 1970s, a few years before Buddy Brennan's visits to Saipan, Sims, began employment on Guam. As part of his business activities, Sims began making weekly trips to Saipan.

Sims soon made friends with numerous families on the island, frequently discussing the disappearance of Amelia Earhart with the older members. He soon noticed that few people were comfortable openly discussing the aviatrix' disappearance.

On one visit, Sims brought his wife with him and as part of a tour of the island, the two were shown Garapan Prison. They were taken to a cell that they were told once held Amelia Earhart. Sims took copious photos of the jail. A few days later, when Sims got the photos back from the processor, he was stunned to see, in one photo of Earhart's cell a ghostly white figure standing in the metal door frame.

Thinking the photo to be some sort of laboratory error, Sims had another print made. It came out exactly the same. In an article in the Kwajalein Hourglass[36], Sims reminisced that as far as he was concerned, the image was a message from Earhart. Had Gene Sims captured the restless spirit of the long-missing aviatrix on film? And what could she have been trying to tell him?

---

[36] Sims,Eugene, "Did Amelia Earhart Land on Kwajalein Atoll?", Kwajalein Hourglass, Vol. 43, No. 2, January 7, 2003

LEGERDEMAIN

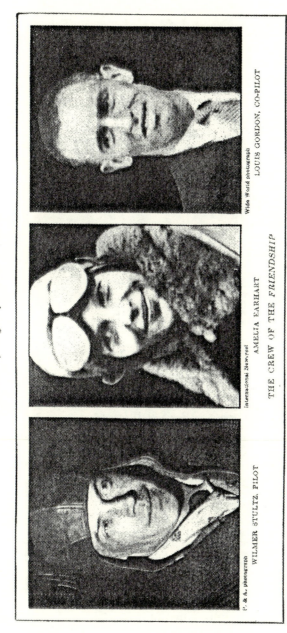

Earhart and her two fellow *Friendship* crewmen
Illustration from the June 30, 1928 issue of Literary Digest

Illustration from Literary Digest of 6-30-28

David K. Bowman

Large Throng Welcomes Amelia on Return to U.S.

Aviatrix Insists Flight Across Atlantic Only "Personal Gesture"

Returns to U.S.

AMELIA EARHART PUTNAM

New York—(AP)—Amelia Earhart Putnam, the girl who didn't want any fuss made over her just because she happened to be the first woman to fly solo across the Atlantic, came home today, still insisting that her flight to Ireland was merely a "personal gesture."

"It meant nothing at all to aviation," she said and added that in her opinion women have been over-praised for their achievements in flying.

"Do you think you've been over-praised?" she was asked.

"Yes," she replied promptly.

Accompanied by her husband, George Palmer Putnam, publisher, Miss Earhart arrived on the liner Ile de France. They were taken off at Quarantine aboard the city yacht Riverside, where a score of widely known fliers waited to greet the tall slim young woman with curly blonde hair who long ago was nicknamed "Lady Lindy."

While tugs whistled, fireboats sent columns of water high into the air, and airplanes swooped and roared, the Riverside proceeded to the Battery, where a crowd of 5,000 or more waited to give Miss Earhart her second welcome home. It was very much like the welcome she received four years ago on her return after her flight to Wales as a passenger with Wilmer Stultz and Lou Gordon.

"I'm just as glad to get back as I was four years ago," she said. "And this time I know the committee better."

One of the first to greet her as she came aboard the Riverside was Elinor Smith who had herself been planning a trans-Atlantic solo flight when Miss Earhart beat her to it.

Someone asked her what she thought about during her night alone over the Atlantic.

"You don't have time to think much—at least not about yourself with relation to your plane," she said. "If you start thinking about yourself, it's fatal."

"I hope the prince was amused," she replied when some one asked her about her meetings with the prince of Wales.

"Like most fliers," she added, "he dances well."

She was greatly impressed by the king and the queen of the Belgians.

"They are both exactly what you'd expect a king and a queen to be like," she said.

Mussolini, she added, showed her a "more gracious side of his personality than some Americans are aware of."

Article originally published in
Appleton Post-Crescent
6-20-32

LEGERDEMAIN

Article from February 1931, newspaper unknown

David K. Bowman

IN THE SPOTLIGHT OF THE WORLD

Here is one of the most recent studio portraits of Amelia Earhart Putnam, the woman of the hour. The famous aviatrix has found herself in the world spotlight as a result of her daring solo flight across the Atlantic, the first ever made by woman. At present Mrs. Putnam is being widely feted in London.

Article from 1932, newspaper unknown

LEGERDEMAIN

Article from 1932, newspaper unknown

David K. Bowman

Amelia Earhart presenting the award at the Earhart Trophy Race at the Cleveland National Air Races in 1935 (NY Times 9-7-35)

LEGERDEMAIN

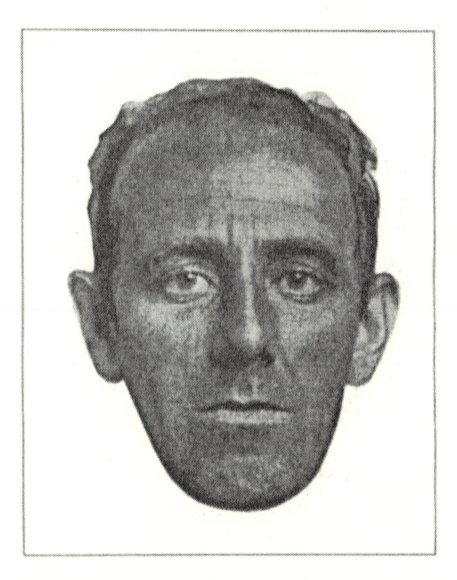

Enlargement of a photo of 'Wilbur Rothar" from the January 1940 Issue of True Detective Magazine. Is this a photo of the man incarcerated as Wilbur Rothar for 24 years?

David K. Bowman

2 news clippings regarding the Rothar Affair (Newark Advocate 8-5-37 and Coshocton Tribune 7-25-37, respectively)

LEGERDEMAIN

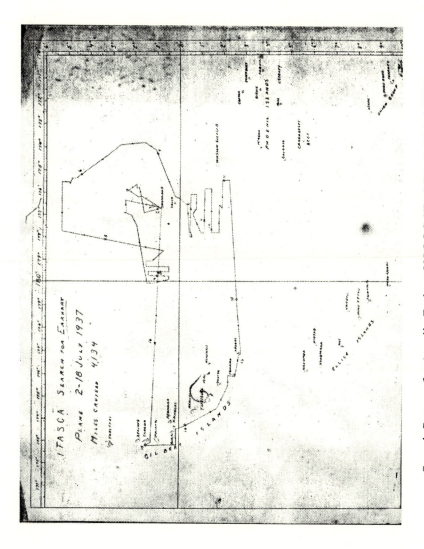

Search Pattern for Amelia Earhart USCGC ITASCA 2-18 Jul 1937

David K. Bowman

65-601-CONFIDENTIAL. 7/19/37.
ITASCA. Radio Transcripts Earhart Flight.

With the above messages the ITASCA's participation in the Earhart flight arrangements, flight, and search can be considered as ended. SWAN and LEXINGTON continued searching for another day and their operations are covered in a separate report.

From the time the Navy took charge of the flight search there is little to report from a communication angle for the reason that frequencies were assigned to the ITASCA and all communications were upon those frequencies. Communication under this system was unsatisfactory and slow. To our knowledge the search under the Navy failed to uncover any new facts and no signals of any sort were heard which might be considered as relating to the Earhart plane.

"SUMMARY OF SEARCH"

1. Earhart plane went down after 0846, July 2nd, and apparently sent no distress message.

2. The conditions of the landing are unknown.

3. From information available until the evening of 5 July it appeared possible for the plane to transmit if down on the water.

4. Lockheed Company on 5 July definitely stated plane could not transmit on water.

5. Amateurs reported several messages, all probably criminally false transmissions.

6. Pan American, Howland and others took bearings on a carrier some place in the Pacific.

7. ITASCA signals calling Earhart, the March of Time program and other signals were iterpreted as from Earhart.

8. If Earhart was down and sending messages the guards maintained by ITASCA, SWAN, SAMOA, HOWLAND, COLORADO, BAKER, PLANE 62C, WAILUPE, PAN AMERICAN, SAN FRANCISCO RADIO, HONOLULU COAST GUARD RADIO and British stations in Gilbert Islands should have intercepted legitimate Earhart traffic, whereas the only interceptions were by amateurs, with the exception of one Wailupe interception.

9. The requests for commercial broadcasts and considerations were a hinderance to the search.

10. Interception of official traffic and release to the public by commercial "scoopers" should be controlled.

11. All available land areas were searched therefore Earhart plane was not on land. Was not heard in Gilberts.

"Summary of Search" by Commanding Officer, USCGC ITASCA
(Page 104)

LEGERDEMAIN

65-601-~~CONFIDENTIAL~~. 7/19/37.
ITASCA. Radio Transcripts Earhart Flight.

12. Extremely doubtful that Earhart ever sent signals after 0846, 2 July.

13. Reports causing diversion of searching vessels should be, and were, carefully investigated. Once the searching vessel receives such a report it is required by public clamor to investigate.

14. The San Francisco and Honolulu monitor systems did excellent work and should be developed permanently.

15. ITASCA's original estimate after three (3) weeks of search problem still appears correct, that plane went down to northwest of Howland.

16. The release of all press details by Headquarters and the Divisions from official despatches is a better solution to handle press than to have searching vessel carrying correspondents whose despatches load up air repeating information already officially given.

It is noted that reference (a) requested a written report of communications throughout the entire expedition with the Commanding Officer's recommendations to be submitted for the information of the Division office.

The ITASCA has been at sea, out of touch with newspapers and commercial radio broadcasting programs. The foregoing report and these recommendations are, therefore, based entirely upon our discussion and study of the matter within the ship. The ship's sole source of information is in the radiograms contained in this report.

The ITASCA has been so close to the matter of the flight and search that it may be this report lacks proper perspective and proportion.

The failure of Earhart to reach Howland and the failure of search efforts to find her was felt by every officer and man on the ITASCA. The ship's company fully appreciated the responsibility of the ship to the Service and to the public.

In the course of time opinions on the Earhart flight and its communications will definitely be formulated. Many of our opinions would probably be changed if Miss Earhart were able to give her side of the picture. It is with this in mind that the foregoing report has been frankly written and it is considered that on this date (July 23) it represents ITASCA thought.

"Summary of Search" by Commanding Officer, USCGC ITASCA
(Page 105)

David K. Bowman

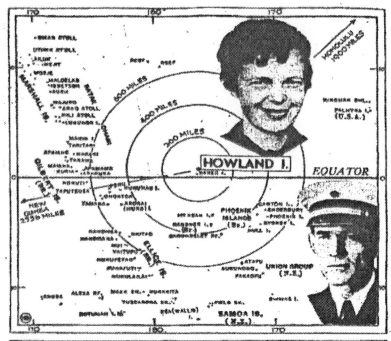

"Where Fate of Flyers is Hidden in Pacific", from Charleston, WV Daily Mail 7-9-37

"Putnam is Kidnaped", Lima, Ohio News
5-13-39

David K. Bowman

A 1936 government photo of a Lockheed Model 12. Note how it is visually identical to a Model 10

LEGERDEMAIN

Lobby card from "Flight for Freedom", RKO Pictures, 1943

David K. Bowman

Amelia Earhart Still Alive Believes Mother; Reaffirms Her Faith

OAKLAND, Cal, April 18— (AP)—Mrs. Amy Otis Earheart, mother of Flier Amelia Earhart, Tuesday reaffirmed her faith her daughter, missing since July, 1937, and declared legally dead, is still alive.

"Amelia was unusually resourceful," Mrs. Earhart said as she hunted for a house in which to make her permanent home near the site of Amelia's takeoff in an attempt to fly around the world. "If anybody could have survived, Amelia would have."

"Amelia Earhart Still Alive Believes Mother",
Nevada State Journal
4-19-39

Chapter 23

The Morgenthau Transcript

One of the most compelling pieces of evidence of government involvement in Amelia Earhart's last flight and her disappearance is a document which has come to be known among researchers as the Morgenthau Transcript. It is a transcript of a phone conversation between Secretary of the Treasury Morgenthau and Malvina Scheider, Eleanor Roosevelt's secretary.

This document was first published in "My Courageous Sister" by Muriel Earhart Morrissey and Carol Osborne. That, in itself, is amazing, since Muriel Morrissey maintained publicly that her sister had not been flying a secret mission for the government. Be that as it may, the document is located in the Morgenthau Collection in the Franklin Roosevelt Presidential Library in Hyde Park, NY. (See copy of transcript and related documents in Appendix VIII.)

Henry Morgenthau was Secretary of the Treasury in 1937 at the time of Earhart's disappearance and one of his duties was oversight of the U.S. Coast Guard.

On April 26, 1938, Paul Mantz, who had been Earhart's technical advisor before her disappearance, wrote a letter to Eleanor Roosevelt, asking her to secure for him the "official report" of the Coast Guard cutter ITASCA in regard to Amelia Earhart's disappearance. Mantz indicated in his letter that he had already tried to get this report by requesting it from the Coast Guard and they had denied his request, indicating that *"the official report (log) could not be released except through certain channels"* (Emphasis is the author's). This last statement seems peculiarly evasive. Why should the government wanted to suppress the search report for a missing private pilot?

What Mantz wanted from the Coast Guard, in actuality, was a copy of the USCGC ITASCA radio log during the period of Earhart's disappearance.

The fact that government records regarding an ostensibly civilian flight were being withheld, speaks volumes and indicates government involvement. Eleanor Roosevelt sent Mantz' letter to Morgenthau on May 10, 1938, along with a short note:

"A little while ago Floyd Odlum and his wife, Jacqueline Cochran, were at the White House when she received the Harmon Trophy for aviation. She told me they all felt that not enough search had been made amongst certain islands where Amelia might be. I told her to send mea memo on the Islands and the reasons why they felt this, and I would transmit it to you and to the Navy Department at once. Now comes this letter which is evidently inspired by Miss Cochran. I don't know whether you can send this man these records, but in any case, I am sending you the letter and let me know whatever your decision may be.

Affectionately E.R".

This short note speaks volumes. It is at once revelatory and cryptic. Not only does it indicate that there are records relating to Earhart's fate which the government is holding, it clearly indicates trepidation on Eleanor's part regarding releasing any of them to Paul Mantz. Mrs. Roosevelt seems to have definitely felt that the requested records would reveal things she would rather not reveal.

An additional document in the file of papers I received from the Franklin D. Roosevelt Library was a single sheet of paper with a few lines near the top, reading:

"6/27
Sent to Mr. Morgenthau – "Can he do this or what do I way [evidently a typo for "say"]? E.R."

"Paul Mantz, Burbank, Calif. Wants Coast Guard data, which was originally turned over to him and he did not copy."

This document, like Eleanor's initial note to Morgenthau, is also revelatory as well as cryptic. "Can he do this or what do I say?" That remark clearly shows strong concern over Mantz' request for the ITASCA Report. And why was the document dated over a month after the Morgenthau Transcript? Was someone just late in making this note?

LEGERDEMAIN

On May 13, 1938, Morgenthau made a telephone call to Malvina Scheider, to discuss Paul Mantz' request for information regarding the Earhart disappearance. The transcript of Morgenthau's end of the conversation remarkably was not removed from the files of material that were sent to the FDR Library in Hyde Park for public access. Morgenthau's staff evidently did not realize the import of the document.

"This letter that Mrs. Roosevelt wrote me about trying to get the report on Amelia Earhart. Now, I've been given a verbal report. If we're going to release this, it's just going to smear the whole reputation of Amelia Earhart, and my . . ." Morgenthau stated after exchanging the usual amenities with Malvina Scheider. He was interrupted at the end of the passage by a remark at the other end of the phone.

"Yes, but I mean if we give it to this man we've got to make it public; we can't let one man see it. And if we every release the report of the ITASCA on Amelia Earhart, any reputation she's got is gone . . ." Morgenthau continued.

"Now I know what Navy did and I know what the ITASCA did and I know how Amelia Earhart absolutely disregarded all orders, and if we ever release this thing, goodbye to Amelia Earhart's reputation," Morgenthau went on. "Now really – because if we give the access to one, we have to give it to all."

Morgenthau then expressed concern over the President's response to anyone in the public questioning the competence of the navy's and coast guard's search for Earhart.

"And we have the report of all those wireless messages and everything else, what that woman – happened to her the last few minutes. I hope I've just got to never make it public . . ." the transcript continued. "Well, still if she [Eleanor] wants it, I'll tell her – I mean what happened. It isn't a very nice story."

Morgenthau concluded his conversation with Malvina Scheider oddly with the remarks: "Well, yes. There isn't anything additional to something like that. You think up a good one. Thank you."

The next day, May 14, 1938, Eleanor wrote a short note to Paul Mantz regarding his request:

"May 14, 1938

My dear Mr. Mantz:

"I have made inquiries about the search which was made for Amelia Earhart and both the President and I are satisfied from the information which we have received that everything possible was doe. We are sure that a very thorough search was made.

"Very sincerely yours,"

This note seems premature, as there is at least no indication from the record that Morgenthau had decided what to do. In fact (see below) it wouldn't be until July 5, 1938 that Morgenthau would announce his decision to Eleanor.

Another interesting and enigmatic document which appears in the file of material I received from the Franklin D. Roosevelt Library is an undated and unsigned note on White House stationery:

"Mr. Morgenthau says that he can't give out any more information than was given to the papers at the time of the search of Amelia Earhart.

"It seems they have confidential information which would completely ruin the reputation of Amelia and which he will tell you personally some time when you wish to hear it.

"He suggests writing this man and telling him that the President is satisfied from his information, and you are too, that everything possible was done".

Since it is unsigned and unaddressed, it is not completely clear to whom the note was written. However a careful reading of it indicates that it is probably to Eleanor Roosevelt from Henry Morgenthau's office. This document underscores the existence of unreleased information regarding Amelia Earhart.

The next document which appears on the record is a short note from Henry Morgenthau to Eleanor Roosevelt. The date "7-5" appears in the upper right hand corner. The note runs:

"Dear Eleanor:

"We have found it possible to send Mr. A. Paul Mantz a copy of the log of the ITASCA, which I think will supply him with all the data he asked for in his letter of June 21st.

"Sincerely,

"Henry"

LEGERDEMAIN

"We have found it possible"? What an interesting statement. It indicates that Morgenthau had perhaps hit upon a way of altering one of the documents in question so that it wold be releasable to the public. This is certainly something that the community of Earhart researchers has been seriously been considering of late.

The record continues with a note dated 7-5-38 from Eleanor to Morgenthau thanking him for finding it possible to send a copy of the log of the ITASCA to Paul Mantz. There seems to be about this short communication a sense of relief. Relief? One wonders why there should have been worry about releasing to the public a copy of a routine document.

Finally, in the packet of papers I received, there is a letter dated July 21, 1938 from RADM R. R. Wesche to Paul Mantz. The letter acknowledges Mantz' request and states that Wesche is pleased to forward the documents. The only puzzling thing is that the letter by Mantz which is referred to is dated June 21, 1938. Curiously, Mantz' letter to Eleanor was dated April 26, 1938 and there was no letter in the packet from the FDR Library dated June 21, 1938. Was the referenced date a typo or was there another letter from Mantz to Eleanor Roosevelt.

The first reaction I had to the Morgenthau Transcript was to Morgenthau's repeated assertions that the "truth" about Amelia Earhart's disappearance would seriously damage her reputation. To what could he have been referring? The apprehension in her voice during her last transmissions?

Another thing is Morgenthau's remark, ".... If we're going to release this, it's just going to smear the whole reputation of Amelia Earhart, and my . . ."

What did the "... and my..." refer to? Is it a subtle hint of trepidation on Morgenthau's part about his and the coast guard's handling of Earhart's support? Was Morgenthau about to say "and my reputation"?

And even more significantly are Morgenthau's remarks about releasing the ITASCA report on Earhart's disappearance. I had reviewed the entire report on a microfilm copy which I secured from the National Archives, several pages of the which are reproduced in the illustration section of this book. The report is the furthest thing in the world from being hugely damaging to Earhart's reputation or memory. It is, in fact, fairly routine.

The inference, then, of Morgenthau's remark about the ITASCA report, is that there must have another, hitherto unrevealed report submitted by the commanding officer of the ITASCA. A report that was regarded as extremely sensitive.

Morgenthau next makes another remarkable statement regarding Earhart disregarding "all orders". This statement, from the lips of the Secretary of the Treasury, unquestionably proves that Earhart was flying for the U. S. Government on her second round-the-world attempt, and that there was more to Earhart's disappearance than was made public.

After some more mundane discussion, Morgenthau then makes yet another startling remark, when he refers to having received "the report of all those wireless messages and everything else". What could he have been referring to? The existing messages transmitted by Earhart that are recorded in the National Archive are relatively sparse and cryptic. The other messages sent between the coast guard, navy and federal government are relatively routine and far from damaging. And what did Morgenthau mean by "and everything else"? Again, we are left with a definite admission by the Secretary of the Treasury that there were additional message traffic and reports, which have never been made public.

Further, there is Morgenthau's remark, "what that woman—happened to her the last few minutes. I hope I've just got to never make it public". What does he mean by it? Make what public? None of the message traffic reflects anything dramatic happening before Earhart disappeared, certainly nothing which would be a disaster if disclosed. Again, there is the clear impression that there was message traffic and other communications that were never released, that there was a cover-up of some sort.

The files from Henry Morgenthau's administration at Treasury have remained closed to scrutiny. Even FOIA[37] lawsuits have not been successful in securing access to these remarkably guarded files, which languish in an impregnable basement vault in Washington, DC.

What grim secrets repose in them?

---

[37] Freedom of Information Act.

Chapter 24

The Eyewitnesses

A pivotal factor in most investigations into the Earhart disappearance is the eyewitness testimony of numerous people living in the Central Pacific area in 1937. It is helpful in some regards, problematic in others.

On the positive side, there is abundant eyewitness testimony to clearly support the proposition that Earhart and Noonan crash landed near Mili Atoll.

One Lotan Jack reminisced to author Buddy Brennan in the 1980s that the Japanese had told him that two fliers had been shot down approximately 30 miles from Mili Atoll and later brought to the harbor on Mili Island.

Tomaki Mayazo, who had been a stevedore in the Harbor at Jaluit in 1937, recalled one evening in July 1937 when he was loading coal on the coal tender KOSHU. A Japanese officer told him to hurry because they [the crew of the KOSHU] had to go and pick up an airplane and some Americans at Mili.

One of the most credible and widely cited accounts is that of businessman Billamon Amaron, since deceased. In 1937, Amaron was a medical corpsman working for the Japanese in Imiej, an island in the Jaluit Atoll. One day in July 1937, Amaron was called down to the harbor to provide medical assistance. He said that the coal tender KOSHU had just arrived from Mili Atoll with a twin-engine airplane hanging in a sling on the fantail.

Japanese told Amaron that they had picked up two people with the airplane, a man, who had been injured, and a woman, who had not. The woman, whom the Japanese called "Meel-ya", had they said, been attempting an around the world flight. Amaron ministered to the man's injuries.

Numerous locals have told researchers such as Joe Gervais, Buddy Brennan and others over the years that the two captured fliers were from Jaluit to Kwajelein, Truk and finally Saipan.

Interestingly, the above information is described by researchers as common knowledge in the Marshall Islands. However, once the captured flier arrived at Saipan, the eyewitness testimony becomes problematic.

The first difficulty is that during early inquiries before 1960, it is reported that almost no-one reported seeing or experiencing anything. Later, during subsequent inquiries, locals suddenly began recalling incidents.

The second, and main difficulty, is that although there are many credible accounts regarding Earhart's and Noonan's capture, there are no unequivocally clear accounts of Earhart's and Noonan's fate. The closest that we come to such an account is the recollection of Nieva Cabrera Blas, recorded by T.C. Brennan in his book "Witness to the Execution". Mrs. Blas clearly recalled seeing what looked like an execution. But when Buddy Brennan excavated the site, albeit some 40-odd years later, he found a blindfold but no body.

Did this mean that the body had already been exhumed by someone else? Or had there been a body there to begin with?

Although there are a lot of rumors of Earhart's and Noonan's executions, there are also rumors of their being spirited away to Japan. Also, some stories were accurate recountings of incidents that actually occurred later.

A good example is a story that Joe Gervais ran down in which a reported crash near Saipan and the subsequent imprisonment of two Americans turned out to have occurred in 1942 rather than 1937. The two Americans were military pilots. This particular incident probably accounted for a number of the eyewitness accounts of Saipan residents.

Another factor to consider is the possibility that the Japanese might have faked an execution to deliberately mislead local residents on Saipan. After all, if they had a pair of prisoners who were political dynamite, the Japanese would have definitely wanted to keep the movements of these prisoners as secret as possible. A good way to accomplish that would have been to make the local populace think the prisoners were dead.

Moreover, disinformation is more than a possibility. In his book, James Donahue asserted that the entire Saipan phase of the Earhart disappearance was a fabrication of the U.S. Navy. He pointed specifically to Josephine Akiyama, an employee of the U.S. Navy as a possible conduit of disinformation.

This author doesn't agree on that point with Mr. Donahue, but feels that disinformation of some sort is not out of the question.

Another writer and researcher, Donald M. Wilson published a seminal work in 1994, "Amelia Earhart: Lost Legend," in which he compiled all of the available eyewitness testimony and even presented at the end of the volume an extensive list of witnesses' names.

Using his considerable database of eyewitness information from Saipan, Mr. Wilson constructed his own explanation for the disappearance of Amelia Earhart. He concluded that Earhart ditched and was picked up by the Japanese, but his conclusion, based on the testimony of resident Saipanese, that Earhart and Noonan were executed later by the Japanese, remains to be corroborated with hard evidence.

So in the end, it is possible to determine from the Marshallese testimony that Earhart and Noonan did indeed come down in their island system and were picked up by the Japanese. It is also possible to determine that the hapless aviators were then shuttled from island to island via seaplane and that the Electra was concurrently hauled to Saipan during this shell game-like evolution.

Eyewitness testimony indicates clearly that Earhart and Noonan were also eventually taken to Saipan. Josephine Blanco Akiyama, previously mentioned in this work, who was living on Saipan at the time, testified that she had seen the plane carrying Earhart and Noonan flying overhear one day in July 1937 towards nearby Tanapag Harbor. Initially Ms. Akiyama reported she thought the plane had crashed. However further years of investigation and rumination by researchers makes it seem more probable that the plane was a Japanese military seaplane and not the Lockheed Electra.

Initially, the two fliers were housed at the Hotel Kobyashi Royokan. Antonio M. Cepeda testified that each day when he went to work, he saw a woman who answered to the description of Earhart outside the building. He said the woman had a thin build, close cropped hair and wore a faded khaki raincoat.

Cepeda described her as calm and unsmiling, and appearing to be depressed, with her thoughts somewhere else. He thought that she was about 35 years old and had heard via rumor that she had been captured while trying to take secret pictures with a hidden camera.

The woman was referred to by everyone on Saipan as "Tokyo Rosa", which mean "American spy lady". [See chapter entitled "Tokyo Rose".]

Mrs. Joaquina Muna Cabrera, interviewed by Fred Goerner in 1962, told of doing laundry for the woman prisoner at the Kobayashi Royokan, characterized as a "hotel for political prisoners". She mentioned that she had only once seen a man who may well have been Fred Noonan in the hotel, with a bandage on his head. She said that the police took him to another place. Testimony of other eyewitnesses indicates that the "other place" was Garapan Prison.

But this is where the testimony becomes slightly confused, as Ms. Cabrera reported that she thought that the American woman died of dysentery at the hotel after some considerable time, perhaps a year.

Where the confusion enters the picture, is that there is eyewitness testimony indicating that Earhart and Noonan were taken together to Garapan Prison where they were kept some time.

Ramon Cabrera, a guard at Garapan Prison, told researcher Joe Gervais in 1960 that he remembered when Earhart and Noonan were brought to the jail, blindfolded and bound. At the time, Cabrera thought Earhart was just an odd-looking, smooth faced man.

Again, there is more confusion as Jesus Salas, a Saipanese incarcerated in Garapan Prison in 1937, was quoted by Don Wilson as asserting that Earhart and Noonan were held there for only a few hours. Buddy Brennan, on the other hand, reported that Salas had seen Earhart in Garapan prison until as late as 1944! [See chapter entitled "Ghosts of Garapan Prison".]

There is further confusion in the testimony, as some of the eyewitnesses indicated Earhart and Noonan were executed while others indicated that they had heard the two had been spirited away.

What was the truth?

It is possible to determine with reasonable certainty that Earhart and Noonan were, for a time, held prisoner on the island of Saipan. But hard evidence of the ultimate fate of the two fliers remains, for the time being, frustratingly beyond our reach.

Chapter 25

Other Strange Reports from World War II

There are a number of other strange experiences reported by participants in WW II in regard to the Earhart mystery.

In February 1944, during an allied invasion of the island of Roi-Namur in Kwajelein Atoll, three marines entered a former Japanese barracks and, during their search of the building, discovered a room outfitted for use by a woman. One of the marines, W.B. Jackson, related that in the room, they found a suitcase containing feminine personal items and a bound, locked book lettered, "10 Year Diary of Amelia Earhart". There was brief discussion, during which two of the marines were in favor of prying open the diary.

Jackson recognized Earhart's name and convinced his fellow marines of the importance of their find. As a result, they later turned the suitcase and it's contents over to an officer at the nearby regimental command post. It was the last they would ever hear of the items.

An intriguing account surfaced in the Kwajalein Hourglass, a publication of the U.S. Army base on Kwajalein. According to the account, during the February 1944 invasion of Roi-Namur, U.S. soldiers discovered a briefcase in the ruins of the island's airport. The bag was embossed with the letters "A.E." in gold leaf. The writer of this article, Eugene Sims, remarked, "For reasons I never understood, the military kept the briefcase find secret until a few years ago [which would have been around 2000, as the Kwajalein Hourglass article appeared in 2003], when it became public and appeared in the Honolulu Advertiser and several other newspapers."

Several significant incidents occurred on Saipan during the invasion of July 1944.

The first involved two U.S. marines, Everett Henson and Billy Burks. In Fred Goerner's book, they reported that one evening after the invasion, a Captain Griswold appeared in their encampment, asking for volunteers for a patrol the next day. He casually mentioned there would be some digging involved in the patrol. Henson and a friend, Burks, volunteered for the patrol, unaware of what lay in store for them.

The next day, Captain Griswold took the two marines to a nearby cemetery, where he checked some landmarks against a map that he had. After some pacing around, they reached a spot at which Griswold ordered them to dig. This was close to a breadfruit tree.

The two marines began to dig. During this process, Henson asked the captain why they were digging. The captain remarked, "Have you ever heard of Amelia Earhart?" Henson answered that he had, and then Griswold said, "I think, then, that's enough said."

Shortly, at about three or four feet, the men uncovered two skeletons, which Griswold ordered them to place in canisters which he had brought. By then, it was early evening and the three returned to the encampment, Griswold admonishing the two against ever revealing the day's activities.

This was an astonishing incident. Did two marines dig up Amelia Earhart and Fred Noonan on a quiet summer afternoon on Saipan in 1944?

Still another incident was later reported by Robert E. Wallack, a marine gunner assigned to D Company, 29th Marines. He and his platoon came ashore near the village of Charan Kanoa and proceeded inland.. Somewhere outside of Garapan City, they came upon a bombed out building. They entered the building and under a pile of rubble, Wallack found a locked safe. The platoon's demolition expert used an explosive to blow open the safe, and what happened after that follows in Wallack's words:

"After the smoke cleared, I grabbed a brown leather attaché case with a large handle and a flip lock. The contents were official-looking papers all concerning Amelia Earhart: maps, permits and reports apparently pertaining to her around the world flight. I wanted to retain this as a souvenir, but my Marine buddies insisted that it may be important and should be turned in. I went down to the beach, where I encountered a naval officer and told of my discovery. He gave me a receipt for the material and said that it would be returned to me if it were not important. I have never seen the material since".

Another briefcase? The above account seems credible enough, but could Earhart have been carrying *two* briefcases on her around the world flight? One wonders.

Another experience was reported by Robert Kinley of Norfolk, VA to researcher Fred Goerner. During the 1944 invasion of Saipan, Kinley's unit, the Second Marine Division, was moving inland from Red Beach One, when they found a house near a cemetery. Kinley, who was a demolition specialist, was sent into the house first to clear the structure of any booby traps. Inside, the young marine found a picture tacked on the wall along with a piece of ribbon. It showed both Amelia Earhart and a Japanese officer standing in an open field, with a background of hills. The officer wore a cloth kepi with a single star in the center of the crown.

A short while later, Kinley was seriously injured in a Japanese mortar attack. He later wrote that after that the photo was gone, either destroyed in the explosion or possibly picked up by some medical personnel.

Kinley also told Goerner that near the house there had been a Japanese command post in a tunnel which Kinley's unit sealed with explosives. Kinley speculated that the officer who had lived in the small house had been trapped in the command center and suggested to Goerner that there may well be other photos or even records in the tunnel.

In yet another incident, one Stanley Serzan, a member of the 4th Marine Division during the Saipan invasion, reported a fellow marine, Sergeant Ralph Cook, found photographs in the wallet of a dead Japanese solder. Cook showed the photos to Serzan, who said one of them showed Amelia Earhart, Fred Noonan and several Japanese officers. According to Serzan, there was another of Earhart wearing a flight jacket of some sort and standing with one Japanese officer. There were also six other photos, Serzan later told Thomas Devine, although he couldn't remember what they showed.

Yet another serviceman, Seabee Joseph Garofalo, came across a dead Japanese soldier on Saipan, and in the soldier's wallet was a picture of Amelia Earhart wearing a short-sleeved shirt and khaki pants. Next to her was a Japanese officer wearing the typical kepi and leggings. Earhart looked very thin and haggard.

According to Garofalo, a short time later, he was involved in the invasion of Tinian, an island near Saipan. In the chaos of battle and the mess of a huge tropical storm, the photo was lost.

In another reported incident, one Sergeant Ralph R. Kanna, of the Intelligence and Reconnaissance Platoon, Headquarters Company, 106th Infantry, 27th Division, was questioning captured Japanese soldiers on Saipan in 1944.

Kanna later reported that one prisoner who was captured in an area called "Tank Valley" had a photograph showing Amelia Earhart standing near Japanese aircraft on an airfield. Thinking that the photo might be important, Kanna routed it to the Army G-2 (Intelligence) Division. Interestingly, when the prisoner was further interrogated by a Nisei, or Japanese-American, interpreter, he stated that the woman in the photo had been taken prisoner along with a male. He indicated that he thought they were subsequently executed.

Evidently, numerous photos of Earhart and her captors turned up during the invasion of Saipan in 1944. Strangely, not one has reached the light of day. At one point, Thomas Devine found a GI who said that he possessed a photograph of Earhart that he had found in a house on Saipan. The man said he had given the photo to a friend of his, who indicated that he had the photo stored in his attic along with several other souvenirs. Subsequently, Mr. Devine began corresponding with the friend, who was being treated in a VA hospital, and offered the ex-GI $10,000 for the photo. Unfortunately, the GI ceased writing to Devine, which put a stop to Devine's efforts.

Is the photo which Thomas Devine once pursued still languishing forgotten in someone's attic? Will one of the serviceman's relatives finally recognize it and sent it to a researcher? And are there other photos out there of the long-missing aviatrix, which now-deceased servicemen brought back with them from a long-ago war, stored away and forgot?

Just one of those photos could go a long way toward solving a mystery.

Chapter 26

Case Closed?

Most investigations into the Earhart affair end up diverging from the official explanation issued by the government in 1937. However, in the 1990's, there were two studies which actually returned to and embraced the official version. This is strange because there has been a large amount of evidence adduced which makes the official version very unlikely.

The first investigation, conducted by Walter Roessler and Leo Gomez, is reminiscent, at least in name, of the noted effort by Gerald Posner, "JFK: Case Closed". The book is a notorious effort by apologist Posner to prop up the long discredited Warren Commission Report.

The approach in Roessler and Gomez' effort, "Amelia Earhart – Case Closed?" must have undoubtedly been inspired in some small way by Posner's work. Their book follows a similar simplified methodology in which only the totally conventional and uncontroversial is accepted. Aside from that, its main weakness is its failure to seriously look at all of the other credible material developed over the years by other investigators.

Nevertheless, "Amelia Earhart – Case Closed?" is certainly much more factual and sensible than the Posner effort. The book is a wonderful overview of many of the technical problems of Earhart's last flight and is recommended reading for that reason. In particular, the section on the crash at Luke Field was very illuminating.

Through excellent research, the authors uncovered what seems to be the actual cause of the crash, which was one of the propellers of Earhart's Electra. Earlier accounts had mentioned the lubrication problem encountered with the propeller on the flight from Oakland to Honolulu, but not gone into it beyond that.

Roessler and Gomez convincingly show that the propeller, damaged by use of incorrect and insufficient lubrication was not fully repairable before Earhart's scheduled takeoff. With no time to send the propeller to a facility where it could be re-machined, Paul Mantz, Earhart's technical advisor, approved stopgap repair work. As a result, the propeller failed again, on takeoff at Luke Field, causing the famous ground loop that the aviatrix experienced.

In addition, the book contributed some great information in the way of detailed lists of all survival equipment and emergency rations taken along on the flight. The lists, shown near the end of the book, are very interesting indeed and make it clear that as long as the craft could stay afloat or ditch on or near land, the hapless fliers should have been able to hold out long enough for help to arrive.

Overall, Roessler and Gomez did a good job of summing up safety, communications and navigational shortcomings of the flight. They concluded that Earhart splashed down approximately 50 miles from Howland Island as a result of a combination of crew fatigue, heavy headwinds, and bad judgement calls re navigation and communication.

The second investigation to embrace the official viewpoint was that of Elgen and Marie Long.

"Amelia Earhart: The Mystery Solved" was a journeyman effort which traced Amelia Earhart's career and then followed her last flight. In trying to fathom and solve the aviatrix' disappearance, Elgen and Marie Long relied heavily on radio traffic records, as well as airspeed and fuel usage computations.

Like Roessler and Gomez, the Longs did not consider much of the material published by researchers in the past 37 years, adhering to the official explanation of Amelia Earhart's disappearance.

Nevertheless, the book is well-researched and contains significant pieces of information.

The first interesting point that I noted the Longs made, somewhere in the middle of the book, was the failure rate of components of the Electra.[38] Amazingly, the Longs reported that after taking off on her round the world flight, Earhart couldn't fly ten hours without experiencing mechanical failures. It was commonplace to have a propeller, electrical or fuel system component fail during a leg of the flight. When she landed at Lae, New Guinea, Earhart had noted that the starboard Cambridge exhaust analyzer was out of commission, which was not unusual, and that there were five other mechanical failures of a non-routine nature.

---

[38] "Elgen Long, and Marie Long, "Amelia Earhart: The Mystery Solved" (NY: Simon and Schuster, 1999) page 175.

LEGERDEMAIN

This puts Earhart's long flight from Lae to Howland Island in a different perspective, making it clear that mechanical failure almost certainly played a part in Earhart's disappearance.

Another related point the Longs made a few pages later was that after landing at Lae, Earhart told radio operator Harry Balfour that she was noticing a lot of static in her radio and wondered if there was a problem with it[39]. As mentioned elsewhere in this report, it has been known that Earhart had radio trouble during her round the world flight, especially just before her landing at Lae. This bit of information from the Longs helps confirm that Earhart's receiver was having problems by the time she reached Lae.

A final bit of information that the Longs adduced was regarding the transmitter of the ITASCA[40]. Although the ITASCA was moored in Honolulu, the ship maintained regular radio communications with the Commander San Francisco Section.

Evidently the Commander of the San Francisco Section must have noticed problems with radio transmissions from ITASCA, as according to the Longs, a message was sent by the San Francisco Section to Commander Hawaiian Section on June 18, 1937 indicating that ITASCA's transmitter was evidently faulty. The commander of the San Francisco Section directed Commander Hawaiian Section to send a technician aboard the ITASCA to correct the problem before ITASCA was due to leave port for Howland Island.

Warrant Officer Henry M. Anthony was sent to the ITASCA to adjust the radio equipment. With him, WO Anthony brought detailed radio arrangement write-ups which had been prepared in March when Harry Manning had been slated as navigator and radio operator. As an aside, the reader should know that when a radio is set for CW (Continuous Wave) for Morse code, any voice transmission from that set will be accompanied with a constant squealing noise.

WO Anthony purported the radio arrangements to the ITASCA as the latest arrangements. As a result, the ITASCA's transmitters were set to CW for Morse code. Unfortunately, in the interim since the March radio arrangements had been formulated, Harry Manning had dropped out of the flight and been replaced by Fred Noonan. Not only were the current members of the Earhart flight at sea with Morse code, they were expecting the ITASCA to be rigged for voice transmissions.

This state of confusion spelled real trouble for the Earhart flight.

---

[39] Ibid, page 183.
[40] Ibid, page 199.

Chapter 27

Deep Sea Searches

A relatively recent development in efforts to plumb the mystery of Amelia Earhart's disappearance is the deep sea search. Prior to the 1980s and 1990s, such searches didn't occur because they weren't technically possible. After the technology became available, it was still financially infeasible for all but the most well-heeled parties. But finally, by the early 90s, such a search reached the realm of possibility.

Then too, a resurgence of the belief in the official version of the affair has occurred as a result of the book by the Longs and also the book by Walter Roessler and Leo Gomez.

Evidently the two circumstances converged in 1994, before publication of Elgen and Marie Long's "Amelia Earhart—The Mystery Solved", when the Longs attempted to launch a deep-sea search of the area northwest of Howland Island.

Long's financial backer was venture capitalist Dana Timmer, who provided the funds to contract with Williamson and Associates of Seattle, WA. Williamson and Associates are best known for their discovery of the ship Central America in 1986.

One of the richest shipwreck finds ever, the Central America carried a cargo of gold that was worth $500,000 in 1857, when it sunk in a storm off of South Carolina. Of course, the value of the gold in 1986 was staggering, and the company which originally insured the cargo, still in business, stepped forward to claim the cargo. The salvaging of the Central America was covered on television in a memorable program in which the camera panned repeatedly across dazzling piles of gold coins on the bottom, still gleaming as brightly as the day they left the San Francisco Mint.

Williamson and Associates' first task was to define a search area for Earhart's Electra, based on Elgen Long's computations. This done, an area of some 2,000 square feet had been designated for the search. Timmer then chartered a Russian vessel for the actual expedition and for a short while, everything looked good. However, when the Russian ship put into Seattle, it was found to be completely in adequate for the task, which caused the entire expedition to collapse badly in a welter of anger and conflict.

Four years later, Long formed an association with another firm, Meridian Sciences of Annapolis, Maryland. Later renamed Nauticos, the company was engaged to locate investors for an expedition to locate Earhart's Electra and afterward to conduct a search for Earhart's Electra in the 2000 square mile area designated by Williamson and Associates.

Shortly after the agreement with Nauticos, both the Longs and Nauticos were startled by news releases issued by Williamson and Associates announcing an expedition to search for Earhart's Electra—funded by Dana Timmer, the Longs' backer.

Timmer had continued his relationship with Williamson and Associates and had put together a new search for the Earhart Plane with the help of a Sacramento, CA businessman. The Longs and Nautico, still without financing, had no choice but to sit helplessly by while the new consortium fielded a search using the Longs' research data.

Dubbed "Howland Landing", the new partnership launched an expedition in mid November 1999 from Majuro Atoll to the target area near Howland Island. The expedition, for some reason, drew almost no media attention and kept their search results to themselves. They later announced in news releases that they had found "a couple" of interesting targets and would return to the area in the Spring of 2000 to photograph them. Since then there have been no other expeditions by the Howland Landing group.

Yet another expedition surfaced in 2001 at the hand of New Mexico millionaire Mike Kammerer, who had provided nearly half the financing for TIGHAR's 2001 Niku III Expedition. Discouraged by TIGHAR's failure to find definitive evidence, Kammerer decided that Elgen Long's thesis was more promising. Christened "In Search of Amelia LLC" the company sought and gained the use of the unmanned submarine ARGUS, which had been developed by the U.S. Navy. As of this date, I could find no word on the success or failure of In Search of Amelia LLC.

Funding for the Longs' expedition finally surfaced and an expedition was sent out by Nauticos in the Spring of 2002, which succeeded in searching two thirds of the 2000 square mile target area. Another Nauticos expedition launched in March 2004 and aborted operations after running into trouble. In 2005 Nauticos plans another try to locate the Lockheed Electra.

Currently there are no signs that deep sea searches for Earhart's plane will stop. For interestingly, in spite of strong evidence that the Electra lies buried elsewhere, some people are still convinced that Amelia Earhart, Fred Noonan and their aircraft rest in over two miles of water somewhere at the bottom of the Central Pacific.

Chapter 28

Tighar

In 1988, a new organization called TIGHAR, The International Group for Historic Aircraft Recovery, was formed. Its investigation into the disappearance of Amelia Earhart, known as the "Earhart Project," is TIGHAR's best known project.

Created by Richard "Ric" Gillespie and his wife Pat Thrasher, the organization launched a well-publicized expedition to Nikumaroro Island, which was previously known as Gardner Island in 1937. TIGHAR's premise was a highly plausible and promising one. Existing evidence indicated the possibility that Earhart had continued south of Howland Island and might have ditched upon the shores of Nikumaroro Island.

TIGHAR thought it likely that the sea had claimed the Electra from the tidal area around the island before the aviatrix and her companion could salvage much of their provisions. Subsequently, TIGHAR hypothesized, the flyer and her navigator then perished from thirst and hunger. A definite feature of TIGHAR's theory was that Earhart had not been flying on any mission for the U.S. government and had not been picked up by the Japanese.

Located not far from the equator, inhospitable Nikumaroro has a hellish average year-round temperature of more than 100 degrees Fahrenheit and oppressive humidity. And although it has some coconut trees planted by German colonists in the 1890s, the island has no other source of food and no fresh water. A typical atoll, little more than five miles long, with the *de rigeur* lagoon in its center, Nikumaroro is host to every bug and misery imaginable.

TIGHAR's hypothesis was quite possible indeed. Nikumaroro was and is no place to be a castaway.

A large, well-illustrated article in Life magazine[41] chronicled the first of a number of expeditions to Nikumaroro Island. It reported that the sole from a woman's shoe, vintaged from the 1930s, had been found by the expedition, along with a piece of sheet metal with rivet holes in it. TIGHAR felt strongly that the sheet metal had come from the fuselage of Earhart's plane.

Unfortunately, the intensive analysis that TIGHAR put the artifacts through in the succeeding years since then could not identify any connection with Amelia Earhart. Experts even went to the length of comparing the piece of sheet metal to existing Lockheed Electras, but the rivet holes didn't match.

TIGHAR countered that Earhart had had some unusual modifications done to her plane, and thus the sheet metal might not match any plates on a standard Electra. The experts, however, were unable to substantiate this.

Not long after the initial expedition, a major television network broadcast a documentary on the TIGHAR expedition. The film was a fascinating presentation, sandwiching excellent archival footage of Earhart between segments on the TIGHAR search. It is a must-see for those interested in Amelia Earhart's disappearance. Many of the archival film clips are positively haunting.

Since then, TIGHAR has made several more visits to the little island, which have yielded no further significant evidence.

Although TIGHAR's efforts to find the fate of Amelia Earhart have not been successful, they have been successful in compiling considerable information on Amelia Earhart and her disappearance. Most of it is available on their very interesting and informative website. A recommended item on the site is a video clip of the last known footage of Amelia Earhart, as she took off from Lae, New Guinea on July 2, 1937.

---

[41] Life Magazine, April 1992.

Chapter 29

Rollin Reineck, Forensics, and a Second Mrs. Bolam

In December 2003, Col. Rollin Reineck, USAF (Ret.) published an account of his efforts in unraveling the disappearance of Amelia Earhart.

Having investigated the Earhart disappearance for many years, Reineck was impressed by the book written by Joe Klaas and Joe Gervais, "Amelia Earhart Lives," but had been unable to develop further new information on Irene Bolam.

However, in the fall of 1991, following a news conference on Amelia Earhart in Hawaii, Col. Reineck's home state, Reineck's research was given the boost it needed. It was suggested that Reineck contact a Mrs. Helen Barber, who had some interesting information regarding Amelia Earhart. He subsequently reached Mrs. Barber via telephone and learned that she was a friend of Monsignor James Frances Kelley, who was a neighbor of hers on the island of St. Croix.

According to Mrs. Barber, in 1981 during lunch Monsignor Kelley related to her that he had been assigned by the Vatican to go to Japan to bring Amelia Earhart back to the United States. Kelley had been picked for the assignment because of his background in psychiatry, having studied under Carl Jung. He related that he had boarded Ms. Earhart at his estate in New Jersey and gave her spiritual, emotional and psychological help. Ms. Earhart, Kelley said, absolutely did not want her survival known upon her return to the U.S., but did not indicate her reasons for this. He also said that U.S. Government officials were working with him to supply Earhart with a new identity.

As corroboration for her story, Mrs. Barber referred Reineck to a Mr. Donald DeKoster, an automobile executive, who had lunched with Monsignor Kelley and heard the same story.

Reineck then called Mr. DeKoster on the telephone two days later on September 27, 1991. Mr. DeKoster indicated that he was a neighbor of the Monsignor, a good friend, and had known him for about 13 years. DeKoster indicated that he had spoken with Monsignor Kelley a number of times about Earhart in addition to the luncheon in question.

According to Mr. DeKoster, Kelley had told him that Amelia Earhart had survived the war, but that she did not keep her identity. Her new name was Irene Cragmile. Later, DeKoster said, Irene Cragmile married a man named Guy Bolam (Bolam was a British intelligence operative). Interestingly, Mr. DeKoster related that Kelley had told him Amelia Earhart did not want her survival known partially because she was afraid she would be connected with the "Tokyo Rose" affair. She also indicated that other wartime issues and the issue of her poor health played a part in her need to remain incognito. She did not elaborate on the "other wartime issues".

The Monsignor had in his possession, Mr. DeKoster reminisced, a number of documents relating to Amelia Earhart. Mr. DeKoster stated that the Monsignor had told him and his wife that he was Earhart's psychiatric confidante and that he had been the Father Confessor to a lot of celebrities.

Mr. DeKoster concluded his telephone conversation, telling Reineck:

"We still see him regularly. I pick him up for a special party we go to every year at the yacht club here. We're quite close. I can affirm that according to him he was involved with Earhart's return to the United States after WWII. And he considered Irene Cragmile Bolam and Amelia Earhart as one and the same person".[42]

A week later, Reineck reached Monsignor Kelley at his home in Rumson, New Jersey and spoke with him over the phone. Monsignor Kelley confirmed that he had been involved in returning Amelia Earhart back to the United States following the war.

An appointment was immediately made for Reineck to meet with Monsignor Kelley at his home in New Jersey a short time later. Two days before the meeting, Reineck called the Monsignor and spoke briefly with him. This time, the Monsignor could not remember much of anything

---

[42] Reineck, pg 187.

about Amelia Earhart. Reineck nevertheless kept his appointment with the Monsignor.

Immediately before Reineck was ushered in to see the Monsignor, the Monsignor's nephew cautioned Reineck that his uncle was 89 years old and sometimes had "bad days" recalling specific events. During the appointment, the Monsignor, although highly talkative and sociable, did not mention or recall anything about Amelia Earhart or Irene Bolam. When Reineck pressed him at the end of the visit, Monsignor Kelley said, "I could not state my feelings. If I were to answer that, I would violate everything I learned in the confessional".

After the Monsignor's death, Reineck obtained permission to examine some of the Monsignor's personal files. He found two enigmatic, but highly suggestive items:

A file folder with the name of Irene Bolam on one side of the tab and the name of Amelia Earhart on the other.

A note was also found inside the cover of "My Courageous Sister" (by Muriel Morrisey) by the Monsignor, reading, "It's too bad that her mother never knew she had survived".

Reineck remarked that although there was no indication that Earhart's mother did not indeed learn of her survival, Muriel Morrissey had indicated in a 1970 letter to Fred Goerner that "Also, I had a *long and revealing conversation with Irene Bolam*" [emphasis that of Col. Reineck]. She didn't say in her letter who had initiated the contact or what was said. Colonel Reineck further observed that if Irene Bolam wasn't in fact Amelia Earhart, Muriel Morrissey would have immediately told Goerner this in her letter. She didn't.

Finally, of the documents which Mr. DeKoster had mentioned the Monsignor had regarding Earhart, Reineck found no trace.

Since the publication of Colonel Reineck's book, more information has surfaced regarding Monsignor Kelley. The information raises almost as many questions as it answers. At least two researchers from the Amelia Earhart Discussion Group, formerly the Amelia Earhart Society, spoke with relatives of Monsignor Kelley on different occasions. In both interviews, it emerged that during his last years, Monsignor Kelley had progressively slipped mentally and become prone to telling some very inventive stories.

One could almost dismiss Monsignor Kelley's account, were it not for the cryptic material found in his files. If the Monsignor had fabricated his account of repatriating Amelia Earhart, why had he left behind the note and the file folder? Had the folder merely held material collected by Monsignor Kelley out of curiosity, as well as facilitate fabrication?

Another indicator of the unplumbed depths of the Bolam affair and the interest it had stimulated came in the fall of 1982, shortly after the death of Irene Bolam. The Woodbridge, New Jersey News Tribune published a series of articles entitled, "Did Amelia die or was she Irene Bolam?" in ten issues between October 18 and 29, 1982. These very interesting articles were printed under the byline of Lois DiTomaso, Sue Emmons, and Donna Kenyon.

Neither completely dismissive of the possibility of Irene Bolam's having been Amelia Earhart nor completely embracing the proposition, the articles went into Earhart's disappearance in great detail.

In some articles, many family and friends of Irene Bolam were interviewed and insisted that Mrs. Bolam couldn't have been Earhart. In another write-up, however, Bolam's doctor, Dr. Man Wah Cheung, was quoted as saying that he remained puzzled about Irene Bolam, even after her death, and wasn't so sure she wasn't Amelia Earhart. In the same write-up, a number of members of the Wings Club, an exclusive aviator's organization in New York City, were quoted as having wondered if Bolam was Earhart, based on her uncanny resemblance to the aviatrix and occasional strange remarks she made.

Another friend of Mrs. Bolam, Diana Dawes, was quoted as referring to Bolam as "a very mysterious lady". Dawes claimed to have a tape recording of one of her conversations and, amazingly, a set of Mrs. Bolam's fingerprints. Mrs. Dawes stated that, "There were many things she said to me that indicated she was Amelia Earhart".

One of the articles reported a remark that Irene Bolam once made which seems a very telling Freudian slip and all but indicates that Bolam was not who she said she was.[43] It was in a discussion Bolam had with John Malloy of Rumson, NJ, whom she met at a golf tournament.

"In all the years I flew, I never wore a parachute", Bolam said to Malloy that day. Since Irene Bolam's entire flight record lasted only six months from late 1932 to the spring of 1933, how could she have flown for "years"? This minor exchange remains enigmatic. Could it have been just a conversational embellishment on the part of Irene Bolam?

A major feature of the articles was a series of photos they carried which were captioned as being of Irene Bolam. Subsequently, screenwriter Tod Swindell commissioned a study which determined that the woman in the photos was not the Irene Bolam whom Joe Gervais met in 1965. In fact, not one of the photos of "Irene Bolam" in the articles appeared to be of

---

[43] "Some Remain Baffled," article, Woodbridge, NJ News Tribune, 10-18-82.

the Irene Bolam Joe Gervais met. The woman had a mild resemblance to the other Bolam, but the two were not the same. Was this a mistake by the newspaper, or was there an agenda afoot?

In 1997, Tod Swindell, a screenwriter and then a member of the Amelia Earhart Society, conceived of the idea of using the then new forensic technique of photographic overlay to determine, once and for all, whether Irene Bolam and Amelia Earhart were one and the same. The technique involved locating photos of the two people involved, which were in similar poses and then overlaying them to evaluate points of similarity in bone structure.

Swindell secured the services of forensic anthropologist Dr. Walter H. Birkby of Arizona, and Dr. Todd W. Fenton of Michigan State University. Amazingly, comparison studies showed that the bone structures of both women, in both face and hands, matched almost exactly.

Although not every pair of photos showed a big resemblance, there were a couple of pairs that exhibited a startling resemblance. Readers should view the photos in Col. Reineck's book to reach their own conclusion.

The two forensics experts reported preliminarily, based on the material that Swindell presented, along with their own comparisons, that there was sufficient scientific basis to conclude that Amelia Earhart and Irene Bolam were the same person. However, as of the publication of this book, the scientists have not issued a final report of the study for scientific peer review.

It is an understatement to say that the Bolam case is controversial. To this day, it is intensely supported by some and hotly rejected by others. We can only hope that the final report will come out soon, which might help to lay to rest a lot of the heated disagreement.

In a 2002 news release, Guy Bolam's surviving brother John related that although he and the rest of the Bolam family had felt that Earhart and Irene Bolam were not one in the same person, the forensic evidence developed by Tod Swindell seemed persuasive to them. They conceded that they didn't think that Irene Bolam was who she appeared to be.

Along the way, however, another mystery surfaced. After a large number of photos of Bolam were secured for the comparison, it was realized that some of the photos were of a woman who *looked* a lot like Irene Bolam, but *wasn't* Irene Bolam. The photos in the News Tribune article, already cited, were of the "Bogus Bolam".

In his book, Reineck referred to the woman whom Joe Gervais had met in Long Island at the early fliers meeting as the "Gervais Irene". The other, spurious, Bolam, he referred to as the "non-Gervais Irene".

Reineck noted that the second Bolam was younger and shorter than Irene Bolam. He conceded that it wasn't clear when the bogus Bolam had surfaced, but made it a fair assumption that she had appeared after Joe Gervais met Irene Bolam at the Sea Spray Inn in 1965 before his talk on Operation Earhart.

Reineck suggested that Earhart was involved in a sort of "government witness protection" program to protect her identity. He suggested that when Gervais recognized Mrs. Bolam as Amelia Earhart, the operators of the protection program responded by producing a second Irene Bolam to throw off anyone who would subsequently pay any undue attention to the enigmatic New Jersey matron. The affair of the Bogus Bolam is, in this writer's opinion, a strong corroboration of Joe Gervais' conclusions.

Was the series of articles in the Woodbridge News Tribune a part of the campaign to discredit Joe Gervais' identification of Irene Bolam?

Ultimately, the appearance of a second Irene Bolam boggles the mind. It was one thing for there to be a cover-up during the politically explosive time of the late 1930s after Earhart's disappearance, and even for some years after. But for it to continue for 67 years, long after the death of Irene Bolam, amazes. What ancient secret would have merited such a concerted and relentless effort to preserve it?

Reineck's book concluded on a haunting note. He had found out that Irene Bolam's brother and sister-in-law had donated an engraved brick to the museum now being operated in Amelia Earhart's birthplace in Atchison, Kansas. The brick was placed in a walkway leading to the front door of Amelia Earhart's childhood home. Engraved on the brick was: "Irene Craigmile Bolam." Further corroboration of the final fate of Amelia Earhart?

Chapter 30

Tokyo Rose

Another strange facet of Amelia Earhart's disappearance was the rumor that arose during WWII regarding "Tokyo Rose". "Tokyo Rose" was the pseudonym of a series of women who made numerous radio broadcasts in the Pacific Theater of Operations during WWII. The broadcasts were designed to disseminate misinformation and demoralize allied troops in the Pacific area.

By 1944, the rumor had surfaced that none other than the missing aviatrix Amelia Earhart was the voice behind the broadcasts. George Palmer Putnam, who had been commissioned a major in the Army Intelligence Corps in 1942, was serving on a bomber base in the Burma area. Thus, when the government decided to send him to monitor the Tokyo Rose broadcasts, it was an easy matter. Putnam was sent to a Marine Corp radio station in a Japanese occupied area of China ostensibly so that he would be as close as possible to the transmission source of the Tokyo Rose broadcasts.

GP's mission was to listen to several Tokyo Rose broadcasts to determine if in fact the voice was that of his former wife Amelia Earhart. After listening to a single broadcast, for less than a minute, GP exclaimed, "I'll stake my life that that is not Amelia's voice. It sounds to me as if the woman might have lived in New York, and of course, she had been fiendishly well coached, but Amelia—never!"

A new perspective regarding GP Putnam's trip to China arose just before this book was completed. As mentioned earlier in this book, I had occasion to meet with Amelia Earhart researcher Ron Bright, and among the subjects discussed was GP Putnam's trip to China. Ron surprised me a little by mentioning to me that his research showed that the only source for the account of GP's trip was Muriel Earhart Morrissey. Thus, he told me, he doubted the provenance of the report and felt that it may never have happened.

Could this well known anecdote be a fiction? That is up to the reader to decide. However, with no solid evidence discrediting the account of GP's trip, this writer sees no reason not to continue to give credence to it.

Another point casting doubt on GP's foray behind the lines in China is that it wasn't really necessary to go to such a dangerous area to listen to the radio broadcasts. In short, GP could have listened to them almost anywhere in the Asian theatre of operations.

A complicating factor was that more than one voice had appeared over radio as Tokyo Rose. Moreover, since there were no recordings made of the broadcasts, it is impossible to say how many different women lent their voices to the effort.

In the summer of 1949, the U.S. decided that a Japanese-American named Iva Ikuko Toguri D'Aquino was Tokyo Rose and they put her on trial in San Francisco. Despite the fact that the defense produced evidence that up to fifteen different women were involved in the broadcasts, the court found D'Aquino guilty on one of the eight counts. She was sentenced to ten years and was released six years later, with time off for good behavior.

Another strange aspect to the Tokyo Rose mystery is that many of the eyewitnesses on Saipan who reported seeing Earhart after her capture, said that residents of the island habitually referred to the woman captive as "Tokyo Rosa". When questioned by Joe Gervais in the 1960s, one Antonio M. Cepada mentioned this fact.

"Why do you call her that?" Gervais queried.

"Everyone on Saipan referred to her as Tokyo Rosa. In 1937 Tokyo Rosa meant American spy lady".

"You mean Tokyo Rose on the Japanese radio during the war? That Tokyo Rose?" Gervais asked in surprise.

"Not that one," Cepada said, shaking his head. "Tokyo Rosa in 1937 meant American spy girl. That's all. Nothing else".

Recent research on this author's part has turned up the repeated assertion in numerous quarters that the name of Tokyo Rose was never actually heard on any of the broadcasts and was a nickname applied to those broadcasts by American GIs.

As a postscript to the affair, Iva Ikuko Toguri D'Aquino was later given a Presidential Pardon by Gerald Ford in the 1970s. There is currently a movement on to ask the Congress to refund to Ms. D'Aquino the $10,000 fine she originally paid the government and award her a Congressional

Pardon. One can only hope that Ms. D"Aquino, who still survives at the age of 88, will receive the long overdue justice she deserves.

It's possible that the name for the Tokyo Rose affair may have had its genesis in the prewar years in the South Pacific, from the slang "Tokyo Rosa" reported by Earhart witnesses. Perhaps GIs who invaded some of the islands picked up that local slang and applied it to the later Japanese radio broadcasts.

Ultimately, the whole affair could be discarded if it were not for several troubling issues. The first is that there are indications that U.S. Army G-2 (Intelligence) seriously thought in 1944 that Amelia Earhart was behind the so-called Tokyo Rose broadcasts. And as we have seen, the government took the rumors seriously enough to send GP into a combat zone to investigate those broadcasts.

Second, Irene Bolam, according to Colonel Reineck, told Monsignor Kelley that a major reason she did not wish to reveal her real identity of Amelia Earhart was that she feared being connected to the Tokyo Rose controversy.

Finally, it seems quite eerie that both the phrase "Tokyo Rose" and "Tokyo Rosa," would have the single common denominator of Amelia Earhart. Is there some further significant aspect to the Tokyo Rose affair, which has yet to be identified, much less plumbed?

Chapter 31

The Carrington Report

One of the most impressively researched efforts is that of Capt. George C. Carrington. Initially entitled "Amelia Earhart: A Report," it went through a number of editions, ending up as "Amelia Earhart: What Really Happened at Howland Island, Unabridged Report IV". According to Carrington, he was the first writer to publish an investigation concluding that Earhart had made an overflight of Truk and Kwajelein Atolls.

Carrington's book is also noteworthy for several other reasons: 1) An excellent discussion and documentation of the political situation in the mid 1930s which surrounded Earhart's global flight; 2) A thorough overview of the reasons for selecting the Lockheed Electra as Earhart's vehicle for her flight, as well its operational capabilities; and 3) A most lucid presentation of the evidence for the government's involvement in the last flight.

A major feature of Carrington's thesis was that what finally caused Earhart's loss was the heavy transmission of "A's" on Earhart's voice frequency, 6210 KC by Chief Bellarts on the ITASCA. Carrington claimed that at the point at which Earhart had neared Howland Island and desperately needed bearings, the ITASCA's transmissions of the letter "A" on her voice frequency prevented her from getting through. Although it is an interesting theory, there doesn't seem to be a lot of evidence to support it. The radio log shows frequent but not non-stop transmission of the letter A in Morse code to Earhart.

One very interesting tidbit which Carrington presented was the report of a strange telegram GP received from someone identifying himself as George T. Huxford. It appeared in the July 8, 1937 issue of the Sun Newspaper. The text follows:

"Amelia landed exhausted in a small boat on a small reef 50 miles west-southwest of Howland Island. She has a weak portable radio, food and water. Hardly the strength to use them. She will be restored alive and taken to a ship, probably Japanese and you will (sic) official confirmation tomorrow. Noonan not with her".

This was a strange communication, which remains unexplained to this day. Carrington feels that it was disinformation generated by the Japanese to distract American attention away from Earhart's actual splashdown further north. But was it? Could it be a hint that Earhart was, at one point, close to being freed?

At the end of his work, Carrington concluded that Earhart had gone down in the Marshall Islands. He not only cited the well-known remark that Admiral Nimitz made to Fred Goerner, but asserted that the admiral made the remark based on discoveries made by the U.S. Navy during the invasion and occupation of the Marshalls and other Central Pacific groups in WW II. This last bit of information is provocative and makes one wonder what else the allied forces may have discovered when they invaded such islands as Saipan, Truk, Kwajelein and Jaluit near the end of the war.

Carrington's discussion of the conception and carrying out of Earhart's mission by the government was detailed, even including the names of both civilian and military personnel involved. Although he did not mention the March Field conference held secretly in April 1937 between Earhart, Bernard Baruch and General Oscar Westover. For some reason, Putnam was not invited and was reported to be very upset when he learned about the meeting following Earhart's return from March Field.

Overall, Carrington's book is an excellent source of information regarding the preparations for Earhart's world flight and leaves us with a much clearer idea of the political and military stewing that was going on during the 1920's and first half of the 1930s. It also presents us, in the form of the Huxford telegram, with another possible piece of the puzzle, tantalizing and cryptic.

Chapter 32

The FBI File

The FBI has made available online copies of documents from their file on Amelia Earhart. It's not possible, of course, to know if it is complete, although some of the documents are quite interesting.

Looking through the copy which David Bellarts kindly provided me, I noted that the first part of the material is comprised of the report of the Navy and Coast Guard search for Earhart, and one on the search conducted by the USS LEXINGTON. The rest of the file contains miscellaneous telegrams and correspondence sent and received by the FBI, several of which are extremely provocative, if not remarkable. None of the following documents have been discussed in print by any other writers.

1943 letter to Walter Winchell:

The FBI file contained a copy of a three page letter to Walter Winchell dated September 22, 1943. As in all the documents in the file, the name of the civilian originator of the document was blacked out.

The letter told a remarkable story. According to the writer, he had been listening to his shortwave radio on the afternoon of July 3, 1937, and at about 2:20 p.m., had picked up a lengthy transmission which he believed to have come from Amelia Earhart.

The writer indicated that the message was picked up accidentally and came in clearly, although he received some subsequent messages, which were overlapped and apparently not as intelligible.

According to the writer, Earhart described [sic] "Mille or Mulgrave Atoll, Klee Passage, Knox Island, and seemed to be located on a small island of 133 acres adjoining Knox Island, directly NE of a part of the Marshall Island [sic]".

Earhart also stated that the provision supply aboard the plane was good, and then described the island where she was, saying that there was no habitation or life, but some vegetation. She said it was a bleak place. She also mentioned an injury to Fred Noonan.

In addition, according to the writer, Earhart described the damage to her airplane and stated that it was drifting. [This is interesting, as Lockheed technicians repeatedly asserted that the radio would not work if the craft were floating in water.]

Finally, the writer indicated that Earhart had spoken the time of her broadcast, using Eastern Standard Time, which had puzzled him, as Earhart most certainly was not down anywhere near the east coast of the United States.

The writer concluded that he was aware that the government was asserting that all messages received by civilian radio operators were found to be false, but that he knew that he was not mistaken about the message he received.

At the end of the letter, the writer thanks Mr. Winchell for his reply to a previous missive and his current interest.

One interesting point is that this account asserts that Earhart was making broadcasts from her aircraft at sea, which although questionable according to Lockheed, tends to lend unexpected credence to the account of Robert Myers.

Also, the area given in the letter for Earhart's ditching agrees with the conclusion of many current investigators.

Given the valid information in the Winchell letter, one wonders if there is something to it. A recent investigation by a Earhart Discussion Group member, however, casts doubt on the authenticity of the letter. Was it written by a crackpot?

1944 and 1945 memos:

The FBI file contained a copy of a remarkable communication written by J. Edgar Hoover, Director of the FBI and addressed to Carter W. Clarke, Assistant Chief of Staff, G-2.

Dated January 18, 1945, the memo reported the experiences of an American soldier in the Philippines during the previous year.

Below follows the text of the memo:

"The information set forth below was furnished to this Bureau by a member of the armed forces, whose reliability is unknown, in the latter part of December 1944. This information which concerns the aviatrix Amelia Earhardt [sic], who was reported to have drowned somewhere in the Pacific some years ago in the course of an attempted airplane flight around the world, is being furnished to you and to the Office of Naval Intelligence for information and of possible interest.

"The informant related that he was attached to the American Forces in the Philippine Islands before Pearl Harbor and that on one occasion, he and another American soldier were entertained by some Japanese in a hotel in the Philippine Islands. He described the walls of this hotel as extremely thin, enabling him to overhear a conversation in English between two Japanese to the effect that Amelia Earhart was still alive and being detained at a hotel in Tokyo, Japan.

'Subsequently this soldier was taken prisoner by the Japanese along with other American soldiers on Bataan and experienced numerous hardships as a prisoner of the Japanese. He stated that at one prisoner of war camp he was given the task of typing statements made by American officers to the Japanese intelligence authorities. He said that one day after a number of interviews had been conducted and he was alone in the office with a Japanese intelligence officer, he inquired of this officer if "My cousin, Amelia Earhardt [sic] is still alive?" The informant stated that the Japanese intelligence officer was apparently surprised by his remark but stated he could tell him nothing except 'Don't worry about her well being. She is perfectly all right.'

"The soldier said he remembered the earlier remarks overheard by him in the hotel in the Philippines and those circumstances caused him to inquire of various Japanese guards at several prisoner of war camps from time to time regarding Amelia Earhardt [sic]. He advised that some of the Japanese stated that they did not know anything about her, but that they had heard her over Japanese radio, others that they had seen her in Tokyo and still others that they had heard she was alive and in Tokyo. The informant said his suspicion that Amelia Earhardt [sic] did not drown at sea in her plane was strengthened by the fact that when working around Japanese air strips as a manual laborer, he had observed certain planes containing instruments which he believed were identical with those in Amelia Earhardt's [sic] airplane.

"The soldier furnishing the above information was referred to the Military Intelligence Division, Washington, D.C".

This is an amazing document. If it is genuine, and if the account is true, it substantiates that Earhart indeed survived her ditching, was taken prisoner by the Japanese, spent time in Tokyo, appeared over Japanese radio, and that some of the technology used in her airplane was appropriated by the Japanese.

Initially, I was troubled by the fact that nowhere in the memo is the solider identified. Aside from making verification problematic, it seemed illogical for Hoover to write a memo like this, marked "Confidential By Special Messenger" and not identify the informant.

On the other hand, there is another document in the file, on U.S. Government memo stationery, dated December 1944, which repeats much of the material in the 1945 memo. This document originally contained the name of the informant, but that had been blacked out in the released copy. If the 1944 memo had been an attachment to the 1945 memo, that would explain the lack of an informant's name on the 1945 document.

Another unanswered question was how the informant was able to recognize the instruments in Japanese planes as having come from a Lockheed Electra. And how could he have known if the instruments were copies of standard ones or of special ones that might have been installed in Earhart's ship. Moreover, Lockheed Electras were available on the open market and it is a matter of record that the Japanese did buy a number of them in 1937.

If both memos are genuine, the ramifications are startling. But, frustratingly, we have insufficient data to prove this.

Chapter 33

We Must Be On You But Cannot See You

A major element in the mystery of Amelia Earhart's disappearance is radio—the equipment she had on her flight and the ability she may or may not have had to send or receive messages.

Although this aspect of Earhart's disappearance is technical, it is necessary for the reader to know a little bit about what was going on with Earhart's radio communications, because communications played an important part in the aviatrix' disappearance. I'll therefore go over this aspect as briefly and clearly as possible—to avoid confusing all concerned.

Interestingly, to this day, the exact radio equipment aboard the Electra is in dispute. A major factor is the fact that alternations were carried out on the airplane at numerous locations both during the preparation for the round the world flight and in Miami after the flight had begun.

There exist copies of a memo from Lockheed Aircraft Corporation[44], dated July 30, 1937, which seems to be authoritative for part of the equipment:

One Western Electric 18C 50 watt transmitter, equipped with crystals for three frequencies: 500 KC, 3105 KC, and 6210 KC.

One Western Electric 4-band Type 20B receiver, covering the following frequency range: 200-400 KC, 550-1500 KC, 1500-4000 KC, and 6000-10000 KC. The transmitter was located on the cabin floor behind the pilot's seat.

---

[44] Muriel Morrissey and Carol Osborne, "Amelia, My Courageous Sister,"
(Santa Clara, CA: Osborne Publisher, 1987) pg. 278.

In addition, there was a device called a dynamotor, which was designed to generate high voltage to power the radio system. It was located under the pilot's seat.

W.C. Tinus, radio engineer who designed and installed Earhart's radio equipment, indicated in a 1962 letter[45] that he went over radio operation with Earhart and Noonan in Miami prior to their takeoff. He noted that the Western Electric unit was specifically set up with a three position lever for 500 KC, 3205 KC and 6210 KC. The 500 KC frequency was reserved for emergency transmissions.

Earhart also had aboard her Electra an additional receiver in the form of an advanced, new Bendix Direction Finder (DF). Connected to the direction finder was a loop antenna mounted on the top of the fuselage. One odd fact is that in some photos of Earhart's plane, instead of the loop antenna, there is a "bubble" type device. Later on, that device did not appear in photos of the plane. (See chapter on this subject.)

Strangely, although testing of direction finder equipment was one of the original reasons that the U. S. government had secretly financed the purchase of Earhart's plane in 1936, Earhart and Noonan were given only the briefest of instruction in its operation.

The Electra was also equipped with a 250 foot trailing antenna which was attached to the Western Electric radio transmitter. Without this antenna, according to W.H. "Walt" Grosselfinger, who worked on Earhart's equipment, Earhart could not transmit far on the 500 KC frequency. In fact, the transmission range of the Western Electric radio at 500 KC, without the trailing antenna, was about 39 miles.

Some sources say that Earhart had her trailing antenna shortened at Roosevelt Field on Long Island, complaining the antenna was causing too much of a "drag" effect on the plane. With a shortened antenna, according to Grosselfinger, the Western Electric radio set would have to be readjusted, and if it wasn't, it couldn't transmit properly. Also, even if the antenna was onboard and the radio adjusted properly, it needed to be deployed fully in order for the radio to transmit properly.

Earhart is reported to have left her trailing antenna behind at some point in the around the world flight, due to its weight and difficult of operation. Unfortunately, it's not clear whether she left the trailing antenna behind in Miami, Florida or in Lae, New Guinea. But without the antenna, Earhart's Western Electric radio transmitter was nearly useless in the 500 KC frequency.

---

[45] Ibid., pg. 277.

Another critical error occurred when Earhart left her telegraph key and its accessories in Miami before her takeoff. Without radiotelegraphy capability, the aviatrix faced serious problems on her Lae to Howland hop.

Also, as mentioned in the chapter on Elgen Long's investigation, it now appears that the Coast Guard had their ships set their radio equipment to send and receive Morse code, instead of voice transmissions. This left Earhart in a hell of a fix and undoubtedly accounts for the lack of any significant radio contact between her and ITASCA. Incredibly, whenever ITASCA sent Earhart a voice transmission, Earhart could probably hear little more than a garbled voice half-drowned out by a high squeal.

On top of the foregoing, Barry Balfour noted in a letter to Chief Bellarts[46] that:

"One important factor that you ought to know is that she handed me her radio facility book the morning she left plus a lot of papers and her pistol and ammunition and I did not see the takeoff because I was QRL with VJZ Rabaul [evidently tied up on the radio] at the time . . . Now inside her facility book was all her radiograms concerning her communication arrangements with the ITASCA and suggested frequencies to be used and she could not have remembered all the information that these papers contained; perhaps she left them in the facility book by mistake, so that would account for her not keeping her schedules with you..."

This last item may have been the most serious error Earhart made prior to leaving Lae. It is borne out by the later serious problems with radio communications.

To summarize, Earhart left Lae, New Guinea without her log of communication procedures and radio frequencies. She had almost no ability to transmit at 500 KC, and impaired ability to transmit on 3205 and 6210 KC. There is strong indication that the Western Electric radio failed at some point after takeoff from Lae. 3205 KC was her nighttime frequency and 6210 was used during daytime.

As far as receiving was concerned, her Western Electric receiver worked well in the high frequency range until sometime after she concluded her radio traffic with Harry Balfour. That may have been as much as six

---

[46] Letter from Harry Balfour to Leo G. Bellarts, date not visible in the photocopy, but probably ca. 1969-70, as attested by a date of 10-1-70 on another letter from Balfour to Chief Bellarts.

to eight hours after her takeoff. Evidently problems which Earhart had experienced enroute to Lae with her Western Electric receiver resurfaced near the end of the Lae to Howland hop.

Another critical point is the physical locations of the transmitters. As mentioned above, one was behind the pilot's seat and one was beneath the pilot's seat. The question has been whether Earhart could continue to transmit after ditching at sea. Lockheed technicians reportedly said at the time that ditching would immediately short out the transmitters.

However, according to Robert Myers' account, ping pong balls were placed throughout the aircraft for buoyancy, and that, combined with the transmitters relatively high location inside the fuselage, raises doubt as to whether the transmitters really would short out after a ditching.

If Earhart's radio equipment did not short out, per the above reasoning, and then post loss messages reported by various private radio operators, Robert Myers, and the writer of the letter to Walter Winchell in 1943 (see "The FBI File"), who reported hearing Earhart broadcasting that her ship was floating at sea might be correct.

Earhart's inability to either clearly hear voice transmissions from ITASCA or be clearly heard by ITASCA due to the mix-up in Honolulu brings her communications problems into clearer focus. It explains how ham operators, and Robert Myers, could hear Earhart's voice transmissions, but not the military.

Another significant factor is atmospheric physics. Frederick J. Hooven noted in his report on Amelia Earhart's disappearance the following. "Because they do not follow the earth's curvature as lower frequencies do, it often follows that signals on 3100 and 6200 KC are not heard relatively nearby, but are heard at great distances because the signals are reflected back down toward the earth by the ionosphere."[47]

In a nutshell, the two main frequencies Earhart was using (3105 and 6210 KC) are prone to frequently being bounced by the atmosphere, which can cause distant radio operators to be able to hear a broadcast, while nearby operators hear nothing.

Moreover, frequencies had to be shifted at dawn because another atmospheric phenomena causes radio signals at the frequency used by Earhart during nighttime to be received at a much stronger level than they are. This in turn makes the sender seem closer than they really are.

---

[47] Research paper by Frederick Hooven in June 1982 on the disappearance of Amelia Earhart, Smithsonian National Air & Space Museum Library, Washington, DC.

David K. Bowman

The above would explain many of the seeming contradictions regarding radio activity during the last flight. Slowly, perhaps, some historical fog may be clearing.

Also, as mentioned in the chapter entitled "The ITASCA," at some point late on July 1st or early on July 2nd on Howland Island, the Direction Finder was inadvertently put out of commission. Chief Bellarts discovered this shortly after Earhart's disappearance, when he was checking the device. He found that Radioman Cipriani, innocently through lack of training, moved the hoop on the Direction Finder in a continuous 360^0 rotation, rather than the short back and forth motion for which it was designed.

As a result, as Earhart approached Howland Island, the coast guard direction finder was in all likelihood out of order and unable to track her.

Another problem Earhart faced was unearthed when Col. Rollin Reineck explained in his book, "Amelia Earhart Survived" why Earhart was noted to never have stayed on the air for more ten to fifteen seconds.

What Reineck determined was that during the flight from Oakland to Hawaii on the first round the world attempt, Harry Manning held down the transmit button on the radio for a long period, to get a bearing from Hawaii, and this had the effect of burning out the aircraft's generator.

In short, when the Electra had been outfitted with all sorts of extra equipment, including the Western Electric radio, the Bendix direction finder and the Fairchild surveillance cameras, the electrical system had not been correspondingly beefed up. It couldn't comfortably meet the demand of all the extra equipment.

As a result, it took little to overload the system. And overloading the system could literally be fatal, because that resulted in complete failure of the electrical system, including instruments. Fortunately on the Oakland to Hawaii leg, the Electra was within line of sight of Oahu and the ship was safely piloted in for a landing. Earhart undoubtedly knew this, according to Reineck, and accordingly always limited her time on the air.

Since Earhart was limited to such short transmissions, the ITASCA couldn't get a good bearing on the Electra with her signal.

When one looks at all of the above, the picture that emerges is a recipe for disaster: equipment failure, confusion, inadequate system engineering, and known atmospheric phenomena that were inexplicably ignored. Or where they?

Chapter 34

Secret Cruises

Another side to the disappearance of Amelia Earhart, to which no attention at all has been given by other writers are two yacht cruises which took place between December 1937 and May 1938.

Although they occurred well after the aviatrix' disappearance in July 1937, they are still of interest for what they reveal about the underlying political dynamics still occurring after Earhart's disappearance. They also indicate that the search for Amelia Earhart was probably still underway in the spring of 1938.

The first cruise involved a yacht called the Athene, which was owned by movie director Tay Garnett. Newspapers in late 1938 reported that Garnett had outfitted his 105 foot boat with a complete photographic laboratory with the intention of going on an around-the-world cruise. The cruise was to allow Garnett to shoot scenic movie footage which would be used to spice up rear screen-projected backgrounds in a series of upcoming films.

Although the 1938 newspaper coverage of the cruise consistently indicated that the Athene left on its cruise in February 1938, I located a December 1937 newspaper with a photograph and long caption indicating that the Athene was just about to leave on the voyage.

Interestingly, the write-up did not mention owner Tay Garnett and the photograph showed George Palmer Putnam, movie actress Ione Reed and the captain of the vessel. Another significant point is that the write-up mentioned only some locations in Mexico and Central America and not the around the world trip touted in papers after the ship's return.

The voyage was reported to have ended in March 1938, at which time there was, oddly, little publicity.

Very interestingly, a cruise that did depart for sure in February 1938 was that of the yacht Nourmahal, which was owned by multi-millionaire Vincent Astor. Astor was a good friend of FDR and also a member of FDR's informal network of amateur intelligence collectors (see Chapter 2, "Islands of Mystery"). From existing correspondence we know that the cruise traveled to the Central Pacific area, especially the Marshall Islands.

We also know that Astor was specifically sent by FDR on the cruise to seek out signs of a military buildup by the Japanese. He was to look for any bases, ports, airfields, or fueling facilities in the Marshall Islands and apprise FDR of their existence.

At the conclusion of the cruise, Astor prepared and had hand-delivered a thirteen page letter to FDR detailing his observations during the trip.

On the surface it would appear to have been just another attempt by FDR to find out what the Japanese were up to in their mandated islands. But could there have been another even more covert reason? It is not impossible that Astor was keeping a sharp eye out for Earhart, as he mentions in his letter to FDR that he had investigated the fate of two recent intelligence missions run by the British in the same area in which Earhart disappeared.

For one thing, there are indications that GP Putnam had become frustrated with the navy's handling of the search for his missing wife. For another, the ship GP was reported to have left on for a "photographic cruise" had been, according to press releases the following year, ideally equipped for an espionage mission, with a state of the art photographic laboratory.

Overall, when one looks at the two overlapping cruise schedules of the ships and the ambivalent departure and return dates of the Athene, it becomes clear that there was probably more going on with these two ships than would casually appear.

For one thing, there were at least three different reasons published for the cruise of the Athene. One was the collection of animals for the California Zoological Society, supposedly partially underwritten by GP Putnam. Another was the filming of a Tarzan-type film featuring Ms. Ione Reed in the title role of a female Tarzan. Yet another surfaced after the return of the Athene, and that was the filming of background footage in exotic locales to enhance a series of projected movies. Which was it? Just the fact that there were multiple published reasons for the trip raises suspicion of an intelligence gathering operation.

It has been speculated that GP may have left the Athene at a certain point in the cruise, very possibly the Galapagos Islands off the west coast of Ecuador. There, some researchers speculate, Putnam might later have been picked up by the Nourmahal before it steamed for the Marshalls, its published destination.

The above is at this point just a guess by some long-time researchers into Earhart's disappearance. However, the timing of the Nourmahal's cruise seems far from pure chance, coming just months after Earhart's disappearance and in light of GP's impatience with the government's failure to locate his wife. Moreover, news releases indicated GP Putnam was listed as "leader" of the cruise of the Athene until it reached Mexico. There, Tay Garnett, the owner of the Athene was scheduled to join the vessel. These circumstances alone suggest there was another covert agenda afoot for the Athene.

What is known for sure is that Vincent Astor, who had at first been given the permission of the Japanese government to visit the Marshall Islands, was later denied entrance into the Marshalls shortly before he arrived in the area.

Astor, and possibly GP, had to be content to cruise the Gilbert and Ellice Islands. Whatever they may have found was never publicly revealed. So we are left to wonder about two overlapping and peculiar luxury ship cruises that occurred after the disappearance of Amelia Earhart. Cruises whose itineraries were ambivalent, whose true purposes were far from above board. Cruises which may have been intended to field a last ditch search for a missing aviatrix.

Chapter 35

The Putnam Papers

One of the most provocative revelations that Joe Gervais turned up in his investigation was what this writer calls the Putnam papers, since they had been accumulated by GP Putnam.

Originally, one of Gervais' partners, Paul Briand, had written to Margaret Putnam Lewis, GP Putnam's widow (at that point remarried after GP's passing) in a letter dated May 1, 1960, seeking her assistance. Mrs. Lewis had indicated in her response to Briand's letter that she was not interested in assisting the investigation. Then, unexpectedly in 1966, she wrote Gervais a letter. It was a bombshell:

"In my silence regarding the Earhart material you were hoping for, I did not purposely wish to deal you such a blow, nor do I want the years to go by without some comment from me on my long silence, but even now though I wish to say something, it is difficult to put into writing the various emotions I have had, and continue to have, over the Earhart material, a part of my inheritance, to which I have theoretically the 'rights'.

"It would surprise you to know what out-and-out STRUGGLES [Mrs. Lewis' emphasis] there have been here on the West Coast to get the material away from me, and you would also be surprised, I think, to know that I have had three different high-priced attorneys helping me protect and defend the material until such time as it can be used as it should be used".

Mrs. Lewis concluded her letter with the astonishing remark, "the ramifications of this whole Earhart chapter of history have been more than I could successfully contend with and even now I am in an uneasy calm awaiting another gale from an unscrupulous source".

LEGERDEMAIN

For some reason, Klaas and Gervais presented Mrs. Lewis' letter without comment or elaboration. It's implications were fairly stunning: GP had accumulated a certain amount of documentation of some sort and it was considered so explosive that repeated hostile legal actions had been mounted to wrest it from Mrs. Lewis. Thus, the Putnam papers may be the single most important discovery of Gervais' and Klaas' investigation, next to Irene Bolam.

What could those papers have contained that a series of major legal battles had occurred over their possession? Mrs. Lewis kept referring to them cryptically as the "Earhart material," giving the impression that they were some sort of momentous collection of material which were damaging to some unknown parties, and were literally a cross for her to bear. And what did Mrs. Lewis mean by "until such time as it can be used as it should be used"? This remark makes it seem as though the revelation of GP's papers could accomplish something worthwhile if disclosed in the right place at the right time. Conversely her remark infers that suppression of the papers might aid and abet some negative agenda.

In 2002, Margaret Putnam Lewis passed away, leaving the Putnam papers to GP's granddaughter, Sally Putnam Chapman. Reportedly, Mrs. Chapman then donated the papers to Purdue University. In light of the disclosures in Margaret Putnam Lewis' letter to Joe Gervais, the fact that there have been no disclosures since Mrs. Lewis' passing leaves one wondering whether *all* of the Putnam papers reached Purdue.

Did Mrs. Lewis, feeling that the time was still not right for disclosure, remove some papers from the collection before her death? Or were the Putnam papers "sanitized" at some other juncture before they reached Purdue University after Mrs. Lewis' death? Or if the Putnam papers did reach Purdue intact, why hasn't some disclosure been made of their contents, if they solved the mystery of Amelia Earhart's disappearance?

Ultimately, all we know is that Mrs. Lewis' letter substantiates that GP Putnam had definitely accumulated some sort of extremely revelatory documents over the years. It is also clear that they were so sensitive and of sufficiently far reaching import that someone with considerable power and money did not want them to see the light of day. They may be history's greatest hope for the truth.

What in the world could have been in those papers?

Chapter 36

What Lies Beneath

There is an aspect of Amelia Earhart's life and disappearance of which most people are unaware. This aspect is almost more interesting than her aviation career and even her disappearance.

For, aside from being the First Lady of Aviation, as she was dubbed, Amelia Earhart was also a considerable psychic. Since she almost never discussed this side of her life publicly, information regarding it is scarce. However, I have uncovered some material which should be placed in this record and which the reader should find interesting.

The lead item in a column entitled "The Washington Merry-Go-Round" by legendary journalist Drew Pearson and his partner, Robert S. Allen, in the February 16, 1937 issue of the Coshocton Tribune, reported as follows:

"One development in connection with the recent air crashes which has been intriguing air officials is the way Amelia Earhart has gone psychic.

"America's foremost woman aviator has now become the No. 1 seeress of the air. She believes she has developed a contact with the occult world by which she knows what happens in air crashes.

"Her latest prediction is that May 10 she will make a startling discovery regarding the crash of the Western Air Express plane lost over the Wasatch Mountains on Dec. 15 between Salt Lake City and Los Angeles, and not yet located.

"Officials at first were inclined to laugh at Miss Earhart's psychic messages. But her accuracy now has them mystified. When a United Airlines plane was lost just outside of Burbank, Calif. Dec. 27, Miss

Earhart called the United Airlines office and told them to look on a hill near Saugus, a little town north of Burbank.

"There the wreckage was found.

"Again when the Western Air Express plane carrying Mr. and Mrs. Martin Johnson crashed Jan. 12, Miss Earhart reported the plane to be near Newhall, 15 miles north of Burbank, where it was found.

"In the earlier crash of the Western Air Express in Utah, Miss Earhart had a vision to the effect that the bodies of the dead had been robbed by a trapper. Two days later, a trapper near Salt Lake City reported finding the wreckage, but then suddenly disappeared without giving the location of the plane.

"This is the unfound plane regarding which Miss Earhart expects to make a startling discovery on May 10".

Two days later in the Elyria Ohio Chronicle-Telegram, for February 18, 1937, an interview with Earhart in Cleveland was published. In part of the interview, she commented on the Drew Pearson column. Given the details of the Drew Pearson column, it was quite clear that Earhart was trying to completely downplay her psychic ability. That portion of the interview ran as follows:

"A few days ago a Washington columnist reported Washington agog over Amelia's 'psychic' ability to predict the location of lost planes. Commenting on the story today, Mrs. Putnam said:

"I suppose the story got started when I was searching north of Salt Lake City when all the others were searching south. The reason I went that way was not because I had been visited by spirits, but because that plane had a 50 mile tail wind and I reasoned it had been further along than the others thought".

A check of newspapers for later in the year, revealed news releases in June regarding the missing plane. The Hammond Times[48] for June 8 reported on operations to remove the bodies of the crash victims and retrieve the mail and valuables from the wreckage. Evidently the craft had been discovered within the previous day or so, and interestingly, there was no mention of Amelia Earhart or the trapper who had reported the downed aircraft.

---

[48] Hammond Times, June 8, 1937.

The two events are the only ones of Amelia Earhart's psychic ability to be found in the records so far. They seem to represent the only instances in which Amelia Earhart's psychic side was revealed.

It's understandable why Earhart would have all but debunked herself in the manner she did, as in a conservative era like the 1930s, a celebrity wouldn't want to be thought of as psychic. In those days, to be known as a psychic carried the risk of being thought of as a "kook" and Earhart's ambitions allowed no room for distractions such as this.

A remarkable article, entitled "Is Amelia Earhart Still Alive?" was published in the December 1939 and January 1940 issues of Popular Aviation. It described the communications from psychics received by GP, and also gave further hints regarding Earhart's own psychic side. The relevant portion ran:

"In looking back through the bright pages of Amelia Earhart's adventurous life, George Putnam remembered something that might explain the curious fervor of all those men and women who wanted to help in his hour of despair. It was simply that Amelia Earhart herself had a fragile psychic quality, some strange susceptibility to conditions beyond understanding. She rarely mentioned it to friends, never discussed it publicly. But whenever AE participated in mental telepathy or psychic experiments to further her curiosity, observers were astonished at the results. And yet she never involved or followed the advice of countless clairvoyants and astrologers who besieged her at every stage of her great flights.

"She used to say, laughing gaily:

"I haven't the courage to tell people my plans in advance. A pilot shouldn't worry and if I listened to every prediction I'd probably never leave the ground".

And, indeed, Earhart received profuse advice from psychics before her various flights. The feedback was usually in the form of warnings to her not to attempt her flight. And of course, each flight had been successful after all. That's probably the reason that, despite her knowledge of the reality of psychic phenomenon and her own psychic ability, Earhart disregarded all psychic warnings just before her round the world flight.

After Earhart's disappearance, psychics entered the picture again, via huge amounts of telegrams, letters and phone calls to GP. Commendably, GP gave impartial consideration to every communication, even occasionally spending considerable sums to follow out promising leads.

Of all the strange communications that Putnam received after his wife's disappearance, the telegram he received on a late July morning at his Hollywood home, in 1937, may have been strangest. It ran:

"Amelia Earhart alive on coral shoal on one of Gilbert Islands latitude 2 above equator 174 longitude. This message received by Mr. L—— New York Medium".

An hour later, GP received a short note from a Captain T__ M__ of Cape Breton, Nova Scotia:
"...I am the retired captain of a copra boat that used to trade in the South Seas. I just happened to remember an uncharted island that we frequently visited for turtle eggs. The Gilbertese natives know where it is, too. The island is at—"

Here Putnam stopped in surprise and called his son David to locate the telegram he had received earlier that day. A few moments later, David returned with the telegram and the two compared the position given in that document with the position given in the note from Nova Scotia. Amazingly both positions were the same!

Putnam called the captain in Nova Scotia and the medium in New York for additional details. The details seemed sufficiently promising that the publisher immediately left for New York City, where two days later, he was able to arrange a check of the island by British authorities. A ship dispatched from Makin Island a short while later steamed for the location given in the two communications to GP.

But eerily, there was no island at the given location. This baffled Captain M— and his former crewmembers, who all swore they had visited the island a half a dozen times. The only explanation that could be put forth was that the island was suddenly sunk by volcanic activity. GP spent $1000 on that adventure.

Probably the most famous psychic incident involved Jackie Cochran, one of Amelia Earhart's closest friends. After Earhart's disappearance, Cochran contacted GP, telling him that she had received strong psychic impressions that Earhart was floating at sea at a particular location east of Howland Island. Putnam practically moved heaven and earth to get the navy and coast guard to search that location. But the search was, unfortunately, fruitless, and Cochran later told GP that it was too late, that AE had perished.

Or had she?

Chapter 37

Dark Agenda

During the mid 1930s, the Japanese were extremely consterned over the Trans Pacific flights inaugurated by Pan American Airways in 1935. The regular commercial flights near Japanese mandate islands represented an almost intolerable threat to the secrecy of illegally constructed military facilities. The Japanese were also quite worried about the U.S. or any other country managing to make hydrographic charts of the Central Pacific. Hydrographic charts, that is, maps showing ocean depths and characteristics, were and are critical in planning military activity.

The Japanese openly voiced their displesure in American newspapers under the headline, "JAPAN IS OPPOSED TO PACIFIC LINE" on March 14, 1935. The article quoted a Japanese Navy spokesperson who had lodged an objection to the pending Pan American trans-Pacific service. He complained that "the distinction between commercial airports and those for naval and military use was not clear. The landing places sanctioned could be converted into naval bases whenever necessary..."

The following autumn, in November 1935, just before Pan American launched the maiden flight of their trans-Pacific run, two Japanese nationals were arrested aboard the China Clipper, caught in the act of tampering with the ship's direction finder.

Several days later, on the third day of the first China Clipper flight, two Japanese fighters forced down a CNAC (China National Aviation Corporation) plane, which was enroute from Hong Kong to Tiensin. As Pan American almost entirely owned the CNAC, this could only be interpreted as an attack on Pan American.

The next instance of hostile disruption occurred on January 5, 1936, when the China Clipper had a portion of its fuselage slashed, during an attempted takeoff from Alameda, California. Someone had placed a slab of concrete with iron rods in it in the takeoff area. The iron rods protruded just enough to damage the passing fuselages of any hapless craft in the area. This incident had all the earmarks of Japanese sabotage.

If the Japanese were consterned in 1935, by 1937 they were almost beside themselves with the series of routes being flown in the Central Pacific, bearing such colorful names as "China Clipper," "Hawaii Clipper" and "Samoan Clipper". The planes being used were technically sophisticated seaplanes, dubbed "flying boats" to bolster public confidence in them.

Shortly after Earhart's disappearance there began a pattern of incidents which indicate that the Japanese, growing hugely restive, had embarked on a quiet unofficial war on Pan American Airways and the United States.

On December 11, 1937, the USS PANAY, a navy gunboat, arrived at Nanking to evacuate Americans from the beseiged city. The next day, December 12, 1937, Japanese fighters attacked the PANAY, sinking the vesssel and killing three persons. The surivors were picked up by several other ships in the area and word swiftly passed about the incident. There was a huge outcry and immense political pressure on Japan. The beleagured country later apologized to the United States making a payment of just over $2.2 million in reparation. Despite Japan's payment, the incident remained infamous.

On January 11, 1938, the next hostile Japanese action occurred when the Pan American Samoan Clipper, a Sikorsky S-42, reported they had jettisoned all of their excess fuel and were aborting their mail flight to New Zealand. At that moment, the Samoan Clipper was in the air not far from Pago Pago, American Samoa. Four hours later, the news came through that the Samoan Clipper had apparently exploded in midair. It was announced later in the year that the Samoan Clipper had exploded due to a quirk in the fuel jettisoning process in which some of the fuel had flowed back into the engine exhausts and been ignited.

However, a big problem with the official story was that no human remains were ever found at the explosion site. None at all. This has suggested strongly to historians that the Japanese had highjacked the craft, took the crewmembers off the plane and then exploded it deliberately.

Six months after the Samoan Clipper disaster, in July 1938 the sensational loss of the Hawaii Clipper occurred. Initially, just as in the

---

[49] Charles N. Hill, "Fix on the Rising Sun," (Bloomington, IN: Author House, 2000)

disappearance of Amelia Earhart, the authorities told the public via press releases that the Hawaii Clipper had been lost at sea. However, years later, just as in the Earhart disappearance, evidence surfaced regarding a Japanese involvement in the loss.

Charles N. Hill, in his recent book[49] on the disappearance of the Hawaii Clipper, detailed how researcher Joe Gervais had discovered what apparently happened to the flying boat.

Gervais made a 1964 visit to Dublon Island in the Truk Atoll as a part of Operation Earhart. His guide informed Gervais that he had been involved in burying the occupants of the Hawaii Clipper on Dublon Island in the summer of 1938. Gervais' guide was a local named Robert Nauroon, who had done contract work for the Japanese before WW II.

According to Nauroon, he and another contractor, Taro Mori, had been hired by the Japanese to pour a foundation for a new hospital for the Japanese Imperial Navy's Fourth Fleet. The Japanese were preparing for a coming war with the United States. When Nauroon, Mori and their crew arrived at the site on Dublon Island, they were confronted by the sight of fifteen bodies, all male, lying face down on the ground. Nauroon noted that some bodies wore civilian clothing and the rest wore dark blue uniforms, reminiscent of the uniforms worn by Pan American Airways crews.

They were instructed to bury the bodies in the foundation they were building and they complied, working extremely rapidly. Both the heat and their grim task made the job quite unpleasant.

Once the job was finished, everyone left immediately, not wanting their Japanese overseers to have any second thoughts about possibly eliminating them.

Many years later, in 1980, Joe Gervais published his discoveries regarding the Hawaii Clipper, but was unable to generate much public interest. Disappointed, Gervais turned his attention back to the Earhart case and that was that. Until the Hill book was published recently.

Hill's book went further than the Hawaii Clipper, looking at a number of incidents during 1937 and 1938. The picture which developed indicated the Japanese government was embarked on a dark agenda of espionage, sabotage, murder, and outright attack against the U.S. A major reason for this activity was Japanese frustration with U.S. trade embargoes against Japan and a desire to disrupt the growing Pan Pacific aviation being conducted by the U.S.

Another factor, in the case of the Hawaii Clipper, was the presence of $3 million in then out of circulation U.S. gold certificates. This money had been collected in the U.S. Chinese community to be sent to the Chinese government as war relief. As well, Japan badly needed an engine design for their war preparations. They greatly coveted the Hawaii Clipper's high tech engines.

Hill convincingly drew a picture of a ruthless campaign, which started as early as November 1935, just before the maiden flight of the China Clipper. The next event in the campaign was the December 1937 attack on the USS Panay in the Yangtze River. Initially, the Japanese released a preposterous cover story, but after a short while they were forced to apologize for the incident, paying the U.S. substantial reparation.

In conclusion, when one looks at the hostile activities occurring in the Central Pacific in 1937-38, the disappearance of Amelia Earhart appears in a new light. Within the above political context, Earhart's having come to grief with the Japanese after having flown near or in Japanese waters is understandable if not expectable. One wonders, in fact, whether Earhart's disappearance was a part of the dark agenda being pursued by the Japanese at the time, which resulted in the loss of two other civilian aircraft in the Pacific.

After all, Earhart's Lockheed Electra was a "state of the art" machine, which contained modifications not available in models for sale in the open commercial market. Those modifications, including a new, classified direction finder, would have been extremely attractive to the Japanese in their war preparations.

Chapter 38

The Real McCoy?

One thing I have noticed in my research and participation in the Earhart Discussion Group, is that the search continues, even after 67 years. It takes on both high budget and low budget forms, from large, well-funded commercial ventures to quiet private digging in various archives.

For the past several years, an outfit called NAUTICOS has been searching the Pacific near Howland Island in an attempt to find the missing Lockheed Electra. The expedition is led by Earhart researcher Elgen Long, who believes that the aviatrix perished, per the official pronouncements, somewhere near Howland Island. The 2004 effort proved unsuccessful, but NAUTICOS, undeterred, will be back out in the Central Pacific in the spring of 2005.

In addition, TIGHAR continues their investigation. They still have hope for their Nikumaroro hypothesis, but they have nevertheless turned their attention to the Marshall Islands.

Also, several new theories have sprung up in recent years. In August 2001, news releases surfaced in the media that an Australian researcher, David Billings had possibly solved the mystery of Amelia Earhart's disappearance.[50]

Mr. Billings had been working with surviving members of a 1945 Australian Army patrol on the island of New Britain. Their experience was interesting indeed. They reported that in May 1945, they were patrolling

---

[50] Zoroya, Gregg "Earhart mystery may finally be solved," USA Today, August 14, 2001.

in the deep jungles of New Britain and trying to keep out of the way of a larger Japanese force somewhere behind them.

Early one morning, point men for the Australian unit came upon the wreckage of a twin-engine airplane. The plane had a fuselage of unpainted aluminum. After brief inspection, one of the men found, on an exposed engine mount, a tag with two numbers on it: "C/N 1055" and "S3H1."

The information was duly noted on a map one man was carrying, a map which now is held by his widow. What is startling is that the construction number or "CN" for Amelia Earhart's plane was 1055. Moreover, "S3HI" was model of Wasp engine installed on the missing aviatrix' Lockheed Electra.

Unfortunately, the numbers held no significance to the patrol members at the time, and with a Japanese patrol in the area, the Australians had to move out immediately

Years later, one of them, Don Angwin, saw a television program about Amelia Earhart, during which photos were shown of Earhart's plane. The photos jogged Angwin's memory and Angwin began to wonder if the aircraft he and his fellow soldiers discovered in the jungles of New Britain in 1945 might have been the missing Lockheed Electra.

Angwin eventually led an expedition of his own to New Britain, which failed to locate the wreckage. Later, news of his expedition came to the attention of David Billings, who is now pursuing an investigation of his own.

The main evidence Billings has right now is the map in the possession of the Angwin's fellow patrol member's widow. Visitors to Billing's website can see a color photo of the map, which is most impressive.

Unfortunately, as plausible as it is, there are major problems with what has now come to be called the New Britain Theory. For one thing, all of the transmissions that Earhart made after leaving Lae, officially acknowledged or otherwise, are in conflict with such a scenario. The content of the last officially acknowledged transmission, that she had half an hour of gas left, as well as the signal strength (level 5) alone make it impossible for the aviatrix to have been anywhere near New Britain.

Additionally, although errors occurred during that fateful flight from Lae to Howland Island, the New Britain scenario would be more than a stretch. It is highly unbelievable that Earhart, with Noonan's navigational skills, became the "Wrong Way Corrigan" of the Pacific, flying nearly back to where she and Noonan had started.

Yet another avenue of research surfaced in 2003 in an interview with an 82 year old WW II vet named Saint John Naftel in the Guam Pacific Daily News. In an article entitled, "It might be the real McCoy,"[51] Naftel told an intriguing story.

He related that while on the Island of Tinian, which is not far from Saipan, he had occasion to work with a group of recently liberated workers. They had been recruited all around the Pacific, including Hawaii to work on the extensive Japanese facilities that were erected on Tinian.

According to Naftel, he began talking with some of the workers who could speak English, and building a dialogue with them as they worked. One day, as they were driving in a truck on the way back to their camp, one of the workers asked Naftel to have the driver slow the truck, saying, "I want to show you something". Naftel complied, asking the driver to slow down when he told him to.

A short while later, the worker told Naftel to slow the truck and as the vehicle slowed, he pointed to the left to two graves.

"Do you see those two graves?" he asked.

"Yeah, I see 'em, so what about 'em?" Naftel said.

"Well, the fifth day I was on the island, they put me on a detail," the laborer began. "They had these two graves and they had these two people—a woman and a man. I recognized who it was because it was that American woman that was going to fly around the world. I heard about it and all before I left the Hawaiian Islands. She had on flight clothes and I could tell.

"[The Japanese soldiers] didn't speak a word, but when we got through buryin' 'em, they came and said, 'If you ever speak about this or mention it and we hear about it, we're going to make you did your own grave and bury you alive. I know they would do it because I had seen 'em do it."

"Holy cow," Naftel said in awed response.

Not long afterward Naftel and the laborer were separated and Naftel never saw him again.

The years then passed until finally Naftel told the story to a retired attorney friend. The friend began writing letters and eventually the story came to the attention of Jennings Bunn, recently retired historian from the U.S. Navy's Marianas Military Museum. Bunn then mounted an expedition that was scheduled to start digging November 12, 2004.

The premise of the Tinian expedition is plausible, but only time will tell. As of this writing, so far, the only word is that nothing has been found on Tinian.

---

[51] Worth, Katie, "It might be the real McCoy," Guam Pacific Daily News, November 8, 2004.

Chapter 39

Legerdemain: Putting Together What Really Happened

A prerequisite to fathoming Amelia Earhart's disappearance is to understand the era in which it occurred. As we have seen in the chapters "Islands of Mystery" and "Dark Agenda", the time in which Earhart went missing was fraught with intrigue, tension and outright hostility. The Japanese were actively trying to harass aviation in the Central Pacific. In the months following Earhart's disappearance, two commercial aircraft were either destroyed or hijacked. Overall, the Central Pacific Area was one dangerous place to fly over in 1937, and in that context, Earhart's disappearance seems much less mysterious.

It is apparent that the "sleight of hand" in Amelia Earhart's final flight, began with the purchase of Earhart's Lockheed Model 10E aircraft. This is shown by the machinations involved in securing the plane in 1936.

Strong evidence exists that money funneled through several covert sources was used to purchase the Model 10E Lockheed Electra, money which originated from the U.S. Treasury. We have the revelation of Professor Potter of Purdue University, candidly admitting that Earhart's plane was bought "in the national interest".

We also have, as mentioned earlier in the book, a document from the FAA archives confirming that a transfer was made of the Lockheed Electra from Lockheed to Earhart for the remarkable sum of $10.00. The likelihood is that the transaction was to legitimize the previous covert purchase of the aircraft.

Not long after taking delivery of the aircraft, Earhart contacted the U.S. Government to ask for their assistance in her round-the-world flight.

It isn't possible to say for sure if she was aware that the government had purchased the craft for her in the first place. Given the indications that GP occasionally made agreements with the government on his own without telling Earhart, it's possible that Earhart was in the dark on the source of the funds for her aircraft.

As previously discussed in this book, FDR, who was already intensely interested in constructing airfields on some of the Line Islands, especially Howland, readily offered assistance to Earhart. The Earhart flight was an irresistible opportunity for the government.

Roosevelt went through the motions of briefly considering and then rejecting the idea of mid-air refueling by U.S. Army Air Force aircraft. He then suggested to Earhart that an airstrip be constructed on Howland Island for Earhart's use in refueling. The situation presented an immediate and perfect pretext to do just that, and in a non-military context.

By the end of 1936, FDR had ordered construction of airfield facilities on Howland Island to commence as soon as possible. During the following month, January 1937, construction was indeed started.

Oddly, although there are records that the government was involved in modifying Earhart's plane during the later part of 1936, the next indication of the hand of the government in the round-the-world flight was after the crash at Luke Field.

At that time, the government reportedly offered to subsidize the cost of the remainder of Earhart's flight in return for full control of the endeavor. At that point, they may or may not have switched Earhart's plane, substituting for her Electra Model 10E an Electra Jr. Model 12A.

News releases immediately went out announcing that Earhart's Lockheed Model 10E was being shipped to the Lockheed plant in Burbank, California for repairs. There may have been a switch in aircraft.

The next thing the government did was to replace all of Earhart's close advisors with people selected by the government. The first and most interesting was Paul Mantz, whom the public was told had simply "stepped down" as technical advisor as a result of the Luke Field crash. Mantz was replaced with Clarence L. (Kelly) Johnson, who later went on to great success as the head of Lockheed's "Skunk Works," in Burbank, California. The Skunk Works developed all sorts of secret, advanced aircraft for the government, including the highly classified Blackbird spy plane and the Stealth Fighter.

Walter McMenamy, head of Earhart's private ham radio network later reminisced in an interview that after the Luke Field crash, all of the cheer and devil-may-care spirit was gone from Earhart. McMenamy, who was never able to speak to Earhart again after the Luke Field crash, reported that Earhart was completely enveloped in a navy entourage and that few of her close friends were able to talk with her afterward. It was very evident to those in the know at that time that the U.S. Government was in complete control of the round-the-world endeavor.

The subterfuge continued at full tilt during the last months of preparations at Oakland Airport. For, as Robert Myers reported, GP handled all visits of official personnel to the airport and may have kept Earhart and Noonan out of the know with regard to many important details.

Putnam had also sold the first publication rights on news releases to the Los Angeles Times. When the government took over the flight after the May 1937 crash, all news releases would be screened by the government and, in some cases, falsified for security reasons. This is what FBI agents told Walter McMenamy, head of Earhart's private radio network in Los Angeles. All of Earhart's radio operators, whose expertise the government badly needed, were sworn to secrecy regarding the operation.

Another telling and incontrovertible point is that the only record the public has of Earhart's final flight, from Oakland, CA to somewhere in the Central Pacific is Earhart's own news releases and reports. In short, the only accounts we have are in Earhart's words. There are no alternate sources of information, other than the official pronouncements, for corroboration. This situation is undoubtedly a major factor that has hampered attempts since 1937 to understand the Earhart mystery.

When the Electra was fully ready for the flight, Earhart took off, eastbound on May 21, 1937. There was no fanfare when she left, as nobody wanted any further embarrassment in case of another accident. The thinking was that if there were any other problems, Earhart could quietly fly back to Oakland for repairs, with no one any the wiser.

There was a brief layover at Tucson, Arizona and on May 22, 1937, Earhart and party took off for Miami, arriving late the same day.

Early on June 1, 1937, Earhart and Noonan took off on the first officially announced leg of the flight, heading south for San Juan, Puerto Rico.

From here on out, until Earhart reached Bandoeng, Java, the round-the-world flight went off fairly smoothly, due partially to GP's organizational work, and partially to the fact that most of the early legs of the flight were relatively short. As a result, Earhart was able to confine most of her flying to daylight hours.

What wasn't mentioned in any of the news releases were the occasional detours off the published flight path for clandestine photos. Thus, her itinerary couldn't be reported in total to the media. News releases had, therefore, to be doctored here and there.

David K. Bowman

It appears that when Earhart reached Bathathia, Java, in mid June, more clandestine activity occurred.

All historical sources merely report that Earhart had laid over in Bandoeng for "an overhaul of her engines". However, this doesn't wash, as the Electra had left on the round-the-world flight in first class condition and its engines had run very dependably. Moreover, Earhart, with her high mechanical aptitude and affinity for her equipment, was well known to take exceptionally good care of her engines. Finally, the Electra had had its scheduled maintenance earlier in the flight, in Karachi. The next major maintenance was scheduled for Lae, New Guinea.

An investigation by Max Clements determined that Earhart picked up a waiting party of mechanics and flew them to Bandoeng, where the Lockheed ship was "souped up" with powerful superchargers. The alternations happened so late in the flight because the superchargers could not be manufactured in time for Earhart's takeoff from Miami. Afterward, Earhart then flew a considerable test flight, some 560 miles, to Surabaya and back, before continuing on to her next stop.

Further indication of the clandestine installation of the superchargers may lie in the fact that Earhart next flew to Kupang, a distance of 1192 miles, in just five hours. That would work out to an airspeed of just over 238 miles per hour, some 88 miles per hour faster that the rated speed of the standard Electra engines.

Between Kupang and Lae, New Guinea, things went fairly well. There was a two day layover in Port Darwin, Australia, that has never satisfactorily been explained. The official news releases of the time explained the layover as being due to bureaucratic delays over medical clearances. However, according to Earhart's notes, made during the layover, the medical clearances were completed quickly and without a hitch. What did Earhart do in Port Darwin for two days?

When at last Earhart took off from Port Darwin, her next and final stop before her disappearance was Lae, New Guinea. It is revealing about the general reliability of equipment in those days that when she landed in Lae, Earhart reported six equipment malfunctions when she landed. This shows another—little considered—hazard in her flight.

She complained to Harry Balfour that her Western Electric radio was giving her trouble. In Lae, too, she experienced a significant delay of three days, although this seemed due mostly to bad weather and scheduled maintenance.

Early in her stay at Lae, Earhart asked Harry Balfour, a radio operator for New Guinea Airways, to fly with her in place of Noonan on the Lae to Howland leg of the flight. When he protested, saying that he would lose his job, Earhart assured him she would see to it he got another job. Earhart evidently realized some of what she was up against on the coming leg of her flight.

Fatefully, Balfour declined Earhart's offer, not telling her that it was because he had a feeling of doom about the flight. He later wondered, after Earhart's disappearance, whether his having flown with her would have changed events for the better. And, indeed, it is very likely that history would probably have been changed if Earhart had taken along an expert radio telegrapher. Many subsequent problems would probably not have occurred.

While she waited for the weather to clear, Earhart evaluated and fine-tuned the plane's cargo. The plane's weight for the Lae takeoff was extremely important, as the runway was short and terminated at a sheer cliff. If the aircraft was too heavy to lift off properly, the aviators risked more than a cheap thrill at the end of the runway.

Fortunately, when the Electra finally took off on the morning of July 2, 1937, it narrowly became airborne at the end of the runway.

But here, over the Western Pacific in the early morning of July 2nd, is where Amelia Earhart's flight reached its moment of reckoning. Ahead lay the longest leg of the round-the-world flight, and huge danger, in the form of violently stormy weather, terrific headwinds, torrential tropical rains, fog, poor visibility, and a Japanese naval force.

As mentioned in the chapter on radio communications, "We Must Be On You But Cannot See You," Earhart had left her notebook containing all radio procedures and frequencies behind at Lae with Harry Balfour. Since she in all likelihood couldn't remember all the frequencies and arrangements, this may well have been a major factor in her disappearance.

After taking off from Lae, Earhart turned to the northeast on a course for either Truk or the Marshall Islands. Evidence from a number of sources points in this direction.

The evidence shows that Earhart made it as far as the Marshall Islands before running dangerously low on fuel, due to the strong headwinds she had encountered. Since it was just after dawn when she sent her last officially acknowledged messages to ITASCA, they were distorted by the well-known physics affecting early morning radio transmissions, causing them to be received at a deceptively high strength on the ITASCA. As a result, ITASCA thought the aviatrix was considerably closer than she actually was.

Per Rollin Reineck, we have persuasive evidence that Earhart turned north after approaching Howland Island. Subsequently, somewhere north of Howland Island, probably near Mili Atoll, according to available evidence, she ditched.

Just before Earhart ditched, she may have been spotted by the Japanese military, and possibly forced down by planes launched by an aircraft carrier.

The Japanese then took their time retrieving Earhart, and wherever she was, she languished for as much as ten days before finally being picked up by a Japanese fishing boat and taken to Mili Island. Later, a ship, in all likelihood the coal tender KOSHU, picked up Earhart and Noonan from the Mili authorities.

On top of the other evidence for Earhart's having ditched in the Marshall Islands, Alfred Capelle, the Marshall Islands ambassador to the United Nations made a statement to the Amelia Earhart Society Symposium in 2002[52]. He was satisfied, from talking with many Marshall Islanders that Earhart and Noonan were picked up by the Japanese in the Marshall Islands and taken to Saipan in a seaplane. A multiplicity of evidence shows Earhart was taken from Jaluit to Kwajalein to Saipan via seaplane.

Ironically, at just about the same time that a Japanese fishing boat picked up Earhart and Noonan, the U. S. Navy sea force arrived to the south to conduct their huge, much vaunted, but ineffectual search. Given the timing, the giant and expensive search was pointless from the start, and, in all likelihood, was put on purely for show by the U.S. government. They had already begun to learn what had happened via communication intelligence and possibly Earhart's own broadcast from Mili.

As mentioned earlier in the book, there is definite indication that FDR, for his own unknown reasons and not any communication from the Japanese, ordered the Navy not to search the Marshall Islands. Why wouldn't he have wanted Earhart to be picked up?

More legerdemain, more sleight of hand. The government already knew they had failed to rescue the aviatrix and the one order of business at that point was predictably to cover up.

After Earhart's disappearance, a veil of mystery descended over events. What is known for sure is that one person found out more than he was supposed to know and was rushed into a mental institution in the fall of 1937 under the name of Wilbur Rothar.

The next documented appearance of the Electra after 1937 was the account of Thomas Devine, buttressed by those of numerous other service members on Saipan. Devine had seen the Electra flying over his camp one

---

[52] Amelia Earhart Society Symposium, 17-21 May 2002, Oakland, CA.

day in the summer of 1944 on Saipan and, later that night inspected the aircraft up close, finding it intact. Still later that night, Devine witnessed its destruction by fire.

The Devine account establishes clearly that Earhart's plane either survived ditching undamaged, or that the Japanese repaired it after its arrival on Saipan. It also establishes that some high official, with authority over the whole island, was on Saipan and orchestrated the destruction Earhart's aircraft. Years later, during Devine's own investigation, he determined that the Electra was destroyed on the orders of FDR.

A masterstroke of legerdemain.

Chapter 40

What Finally Happened to Earhart and Noonan?

As observed thus far, eyewitness accounts of Saipanese and other island locals in the area are helpful but problematic. This is because there is frequently no hard evidence to corroborate whom it was that the eyewitness saw in their experience(s), or even when. Also, some accounts are based on rumors, i.e., "I heard that the two Americans were later executed" or "I later heard that they were later taken to Tokyo".

However, the eyewitness testimony does help to establish that Earhart and Noonan were definitely taken prisoner by the Japanese, somewhere in the Marshall Islands. It is clear that all of those people cannot be completely mistaken. In addition, Admiral Nimitz' revelations in 1966 reinforce this hypothesis.

As to what happened after Earhart and Noonan's capture, there is evidence regarding Earhart's ultimate fate. Several independent and unrelated sources have yielded evidence cross-corroborating the likelihood that Amelia Earhart may well have survived ditching and imprisonment. Most of these sources, Joe Klaas and Joe Gervais, Robert Myers, Rollin Reineck and Tod Swindell have already been discussed in this book. But yet another is David Deal, an artist from California who has done extensive work similar to Tod Swindell on overlays of photos of Amelia Earhart and Irene Bolam.

One of Tod Swindell's overlays, shown on the dust jacket of this book, demonstrates the haunting physical similarity between the two women. Dave Deal's overlays are even more striking. Did Amelia Earhart assume the identity of Irene Bolam?

Readers are invited to turn to Appendix XII, review Dave Deal's overlays and decide for themselves.

This proposition does not seem at all shocking or impossible. Especially in view of the facts already related in the book regarding the Lae to Howland flight and the situation with the Japanese. Moreover, being captured by the Japanese was far from unheard of in that era. Starting as far back as 1923, military officers in search of intelligence came to grief with the Japanese in the same area Earhart flew over. During WWII, numerous people, well-known or otherwise, were captured by the Japanese. Colonel Pappy Boyington for example, as well as General Jonathan Wainwright, ended up as prisoners of the Japanese.

As noted in the chapter entitled "The Bolam Affair," Joe Gervais had encountered Mrs. Bolam socially at a meeting of old time fliers. Gervais' experiences with Mrs. Bolam, reported earlier in this book, were highly suggestive that the wealthy and reclusive matron was far more than she appeared. Even Mrs. Bolam's brother-in-law did not believe she was who she seemed to be. Mrs. Bolam's own words to Gervais over the telephone suggest the truth: "I once had a public life. I once had a career in flying. But I've retired. I've given all that up now".

Robert Myers' account, although quite problematic in some ways, nevertheless offers some mild cross-corroboration for the Gervais account. This is because Gervais only had a few, occasional and enigmatic contacts with Mrs. Bolam. Myers, on the other hand, claimed to have experienced a four year long series of contacts with Mrs. Bolam over the telephone, climaxed by a face to face meeting in 1982. Moreover, Mr. Myers said he recognized Mrs. Bolam as Amelia Earhart even before he knew of her published name.

Even more telling, in his book Robert Myers published a number of photos of Mrs. Bolam, as well as Joe Gervais' photo. The only way that Mr. Myers could have published the non-Gervais photos would have been if Mrs. Bolam had sent them to him, as they all appeared to be personal photos of hers. There were captions explaining where each photo was taken, which could only have come from Mrs. Bolam. These photos alone are more than persuasive. One profile photo in Mr. Myers' book looks eerily like a 79 year old Earhart, even to the 1930's style beach hat she was wearing. In fact, the similarity to Earhart in that particular profile shot seems to go beyond mere chance.

Many media sources cavalierly dismiss Mrs. Bolam as "a New Jersey housewife". Even Irene Bolam herself insisted she was an "ordinary housewife". The fact of the matter is that, whoever she was, Mrs. Bolam was not by any stretch of the imagination a homemaker. Mrs. Bolam was, if anything, quite wealthy and influential, moving quietly between her multiple homes at will. She was noted to be well-traveled, well informed, remarkably well-connected and was anything but a workaday, floor-sweeping, meal-cooking housewife.

For the sake of fairness and balance, there is a definite chance that Amelia Earhart did not assume the identity of Irene Bolam. Irene Bolam might actually have been an "agent of distraction", employed by someone to divert the attention of researchers from the real fate which befell Amelia Earhart. Additionally, it occurred to me just before I finalized the manuscript for this book that Irene Bolam might also have been a bit of a fabricator near the end of her life, like Msgr. Kelley.

It's not impossible that Irene Bolam, who reportedly enjoyed the attention she received, simply took pleasure in "wigging out" friends and sending investigators in excited circles.

However, the circumstantial evidence is strong enough at this date and there is enough cross-corroboration that it appears more than possible that Amelia Earhart might have survived her ordeal in the Pacific, later assuming the identity of Irene Bolam, amazing as it may seem.

As for the fate of Fred Noonan, he may have perished at the hands of the Japanese, or he may have been taken to Tokyo, as some Saipanese witnesses report or speculate. But unfortunately, there is far less information on his fate than on Earhart's.

Both Robert Myers and Joe Gervais in their books indicated suspicion that Fred Noonan and William Van Dusen were one and the person.

In the middle of Myers' book are photos on facing pages. One of them, on the left side, is of Fred Noonan standing with Amelia Earhart in front of the Lockheed Electra in 1937. He is wearing a dark polka-dot tie. On the opposite page, is a photo circa the mid 1970s, of William Van Dusen, who had displayed such an unusual interest in Major Gervais' investigation.

When one compares the two photos, the men's faces appear not to be just similar. They appear to be identical, down to their distinctive sharp noses and the deep creases at either side of their mouths. They're both even wearing what looks enigmatically like the same dark polka-dot ties. William Van Dusen appears for all the world to look like an elderly Fred Noonan.

In Gervais' investigation, as noted earlier in this book, he encountered the enigmatic Mr. Van Dusen and was struck by the physical resemblance between Van Dusen and Noonan. Van Dusen also made it a point to flourish Noonan's cigarette case at dinner one night with Gervais.

Some months before completing this book, I corresponded via email with noted aviatrix Ann Pellegreno, who retraced Earhart's flight in 1967.

Ms. Pellegreno told me that she had acquired Fred Noonan's cigarette case at auction and had also met William Van Dusen before his death. She indicated that, indeed, he was a startling look-alike for Fred Noonan but that she had satisfied herself that he was not in fact Fred Noonan.

However, it occurred to this author that although Ms. Pellegreno established that William Van Dusen was not Noonan, she did not establish that Noonan had not been *posing* as Van Dusen. Ala Irene Bolam.

Still, haunting doubt remains on several counts. First, in a passage from his book, Robert Myers mentions that Noonan, who washed and repaired airplanes at the Oakland Airport just before being hired by Earhart, went not only by the name of "Fred" but the also by the name of "Bill". Minor disagreements even reportedly developed when some of the mechanics argued as to whether Noonan was "Fred " or "Bill". "Which one is it?" one mechanic demanded.

Second, Fred Noonan's cigarette case, which William Van Dusen possessed, presents a troubling puzzle. Noonan would have been expected to have taken the cigarette case with him on the 1937 flight around the world or to have left it with his new wife. If that is so, the possession of the cigarette case 28 years after Noonan's disappearance by a third party is an indicator that the navigator may possibly have returned from his Pacific ordeal.

An additional thought is that the man who purported himself to Joe Gervais as Bill Van Dusen knew Irene Bolam, kept in close contact with her, and evinced extreme interest in Gervais' investigation. This in itself suggests that there was something covert going on in regard to both Bill Van Dusen and Irene Bolam. But what?

Another possible hint as to Noonan's fate turns up in Rollin Reineck's book. On page 76, there is a photograph taken on Saipan, probably in 1944, which shows a group of military personnel walking down a war torn street. It shows both U.S. personnel and Japanese military prisoners. On the right side of the photo, however, is a lone man dressed in a white shirt and civilian type pants. He bears a certain resemblance to Fred Noonan. Unfortunately, while the photo is an interesting bit of information, Reineck presents it without elaboration, leaving us with no way to evaluate the image.

One possible solution, suggested by the facts, is that returning from Japan, Earhart was weak and in poor health mentally and physically. Very afraid of being involved in the Tokyo Rose uproar, she might even have made, purely to survive, one or two radio broadcasts during the war. GP's brief adventure in China, at a remote Marine Corps listening post indicates that the government gave credence to this scenario. GP's posting in the

Burma area makes one wonder, as the publisher was 55 years old when he was commissioned in 1942. The government does not go to the expense and trouble of commissioning a man nearly 60 years of age and sending him halfway around the world without reason.

Furthermore, to Earhart, her disappearance in 1937 may well have represented an embarrassing professional and financial failure that she wanted to forget. She was, after all, a woman who was a perfectionist and accustomed to success. On top of that, the ailing aviatrix may not have wanted to return to a punishing celebrity, relentlessly promoted by GP. Moreover, Myers stated in his book that Earhart/Bolam was still [in the early 1980s] "between Putnam and the government", an intriguing and provocative remark. These reasons may, after all, be the quiet truth.

Did Fred Noonan meet his end at the hands of a Japanese irritated at his ill humor? Or did he manage to survive years of imprisonment, possibly on Saipan, to be freed in the 1944 invasion? And if after returning home, was he able to camouflage himself with the identity of a man known to be almost his twin? And why?

Or is the simple truth that young Bobby Myers, as well as possibly the mechanics at the old Oakland Airport were just taken in by the extraordinary resemblance between Van Dusen and Noonan and they thought they were the same man?

In the end, Fred Noonan remains almost more of a mystery than Amelia Earhart. We are left to ponder the troubling similarities of two photos, the forgotten story of an airport mechanic known by two names, a fellow aviator who appeared to be his exact double, and a silver cigarette case.

Chapter 41

Echoes of a Cover-up

When Amelia Earhart disappeared, the first echoes of a cover-up may have sounded almost immediately in the patterns of the media coverage. Initially, it was reported that Earhart had probably made a safe landing on a small island. SOS messages were reported to have been sent by Earhart using her airplane engines to generate power. GP and Paul Mantz were reported to be convinced that she would soon be located. Two days after Earhart's disappearance, the media had markedly changed their tune, reporting that SOS messages had been received from the aviatrix, indicating that she had ditched at sea.

Within two more days, the media began to report that the limited number of navy ships searching the Howland Island area had found no trace of Earhart or Noonan. Over the next ten days the conclusion reached in the press was that Amelia Earhart had ditched in the sea within 40 miles or so of Howland Island, perishing immediately. This conclusion supposedly came from the U.S. government.

Frequently, the earliest press coverage of an event is usually the more candid and honest. After that, if the event is considered sensitive enough by the establishment, press coverage may be distorted with disinformation and misinformation until the truth is no longer distinguishable. It happens frequently in high-profile events that could embarrass a government or other powerful group.

Could it be that the earliest reports of Earhart's disappearance were the most undistorted and accurate? Could the later changes in coverage be the faint echoes of a cover-up? The fact that no slightest trace of Earhart's Lockheed Electra was found—no oil slick, no debris, ping pong balls—

indicates strongly that the Electra did not just pile into the sea. Something else almost certainly occurred. Airships that pile into the sea leave traces.

Another eyewitness account surfaced recently, which strongly buttresses the thesis that Earhart was on a classified mission for the government. Alex Coutts, a navy mechanic who was present at Luke Field in March 1937 recently reminisced that after the crash, some of the naval personnel at the field saw surveillance cameras installed in the Lockheed Electra.

According to Coutts, "Everyone on the base knew about the photographic equipment aboard the Electra—it was the talk of the base." Even more surprisingly, Coutts recounted that it was "common knowledge" in the military that Amelia Earhart was on a "mission" for the government.[53]

From 1937 until the final year of WWII, as far as the public knew, their beloved aviatrix had gone to a watery grave without a trace.

Before U.S. troops began taking Japanese strongholds in the Central Pacific, there were only rumors among service members and among civilians in the U.S. about the vanished flyer. In early 1944, the Tokyo Rose Affair added fuel to the fires of rumor, climaxing in the sending of GP Putnam to China to monitor radio broadcasts.

By late June 1944, American forces were invading and securing previously Japanese-controlled islands such as Saipan and Jaluit and discovering things. As has been shown in this book, there is credible evidence that Earhart's Electra was found in a hangar at Aslito Field on Saipan. Servicemen such as Thomas Devine saw the old plane in a hangar, in the air and/or being destroyed by fire on the landing strip. They discovered that there was some covert agenda in operation, and were puzzled by the government's secretiveness regarding the aging Electra in an abandoned hangar. Further echoes of a cover-up.

In recent years, researchers have unearthed definite evidence that, sometime after his assignment to China to listen to Tokyo Rose broadcasts and before he was sent home in December 1944, GP visited Saipan. This was apparently not long after the invasion of Saipan. GP secured a jeep driver and traveled around the island interviewing locals in regard to his late wife's disappearance.

In light of the Devine account and its timing, there is a clear inference that GP's visit to Saipan may have been due to his having learned of the recent discovery of the Electra at Aslito field. This also further buttresses

---

[53] 2004 Symposium of Amelia Earhart Discussion Group, 8-21-04, Oakland, CA.

the Devine account and the proposition that Earhart ended up on Saipan. Another echo of a cover-up?

Evidently discoveries on Saipan in the summer of 1944 somehow filtered out of the war zone to the United States to spark some interest in the media.[54] In the fall of 1944, an article appeared in the American Weekly entitled "Amelia Earhart: How Long a Mystery?" detailing some discoveries in the Pacific area by U.S. service members. Unfortunately, one of the discoveries referenced in the article was one that I have not been able to identify so far.

The article referred to an album of photos of Amelia Earhart being discovered by an American serviceman in an abandoned barracks. Oddly, there are no reports that an album of photos of Earhart was discovered during the invasion of Saipan. Was this report perhaps a confused version of the report of the discovery of a photo of Earhart discovered in the quarters of a Japanese officer on Saipan? Or was there an album discovered on Saipan that has gone unreported until now?

Aside from that problem, the article focused heavily on the experiences of Dr. M. L. Brittain, president of Georgia Tech. Brittain had been on the USS COLORADO in 1937 when it was pressed into the search for Amelia Earhart. Brittain related that during the 1937 cruise, he got the very definite feeling from the talk among the ship's officers that Earhart had indeed been involved with an intelligence mission for the government on her last flight.

Brittain further speculated that it was possible that the aviatrix was still alive, in the hands of the Japanese and might possibly be rescued one day in the not too distant future.

The article also pointed out that in February of that year (1944), U.S. Representative James Heffernan had called for a new army-navy search for Earhart among the Japanese-controlled islands in the Central Pacific. Moreover, the article cited the heavy interest of the British, New Zealand and Australian governments in Earhart's disappearance as proof of the proposition that the aviatrix was involved in some covert mission for the U.S. government.

The article next tantalizingly referred to some "confidential files" of the ITASCA, which indicated that Earhart's communications did not comply with the "pre-arranged plan." What files could the American Weekly writer have been referring to? What communications could the

---

[54] American Weekly, 9-10-44.

writer have been referring to? Could this be another echo of a cover-up?

The article concluded with mention of rumors picked up by U.S. servicemen in the Pacific area indicating Earhart had been taken to Japan. Overall, the 1944 American Weekly article is remarkable for its early references to Earhart's possibly having crashed near Jaluit in the Marshall Islands and having been taken prisoner by the Japanese.

Despite the flurry of attention to the Earhart affair, things quieted down again until the 1960s, when a remarkable spate of investigations popped up and books began to be published. Another echo of a cover-up may have surfaced during an interview that newsman Fred Goerner conducted with Amelia Earhart's former secretary, probably Margot DeCarie, who asked for anonymity in Goerner's book. Goerner started the interview with a quick series of questions.

"First, do you really think Purdue University bought that plane [Earhart's Electra] and do you think that it was intended for some kind of vague experimentation? Second, if the whole thing was a publicity stunt as a lot of people seem to think, why did the government assign some of its top experts to the flight and why did President Roosevelt have an airfield built for her? Last, do you believe the President ordered the Navy to spend $4 on a search for a couple of stunt fliers?"

"Won't you tell me what you think?" Goerner then asked Earhart's former secretary.

Startlingly, the secretary said, "Only this. President Roosevelt knew about everything. He knew the price Amelia paid."

"Don't you feel it's about time Amelia received some justice," Goerner asked.

"When one does the things Amelia was doing, one can't expect to receive justice. She knew that. She talked to me."

"Do you think there's any possibility Amelia is still alive?" Goerner then asked.

"She's dead. She died a long time ago. If she had survived the war, she would have come home even if she had to swim," the secretary said.

"Do you think the Japanese captured her?" Goerner continued.

"Of course they did," the secretary said flatly.

This was an amazing interchange indeed. Earhart's own secretary candidly inferred that Earhart had taken on a risky mission for the government, that FDR knew all about the aviatrix' fate, and that Earhart had most certainly been picked up by the Japanese.[55] As mentioned earlier in this book, DeCarie also spoke to Mr. and Mrs. Don Kothera, reconfirming her feeling that Earhart had been on a mission for the government. She also revealed that she had destroyed an envelope full of documents for the aviatrix, when she failed to return from her last flight.

Simply amazing. This wasn't just an echo of a cover-up, it was almost a crash.

Goerner also reported interviewing Fred Noonan's widow, who had since remarried. The first thing Mrs. Ireland did was to mention the old rumors that Earhart and Noonan had developed a romantic relationship during their association. She showed Goerner several love letters written by Fred Noonan during the round the world flight. It was immediately obvious to the newsman that the letters were from a man completely in love with his wife. Any mention of Earhart was innocuous and distant.

Near the end of the conversation, Mrs. Ireland told Goerner that she thought there were things that Fred had not told her about the flight around the world. An aura of secrecy had surrounded the flight. She said that when she had questioned Noonan about the flight and especially the route across the Pacific Ocean, he became evasive.

Further, she said that following Earhart and Noonan's disappearance, the official attitude in Washington, D.C. took on a disturbing aspect to her. "It was as if no one ever wanted to hear the names Noonan and Earhart. No one was supposed to dispute the decision they were lost at sea."[56]

Another amazing revelation. It seems to document secrecy, a covert mission, and a subsequent cover-up.

It became glaringly apparent to Thomas Devine when he found he had been tailed by ONI agents on a visit to the home of Muriel Morrissey, that there was definitely a government agenda afoot. Someone was hugely worried about all of the interest in Amelia Earhart.

During the same period of his visit with Muriel Morrissey, Devine, as related earlier in this book made a full report of his Saipan experiences to the Office of Naval Intelligence (ONI). Devine had some photographs at home that he wanted to send to the navy later, and was told, surprisingly, not to send the photos to the ONI office in Hartford. Instead, he was told to send them to another agency, the Security Risk Agency (SRA) at the Hartford, Connecticut Naval and Marine Corps Reserve Training Center.

Security Risk Agency? How could the 23 year old disappearance of Amelia Earhart constitute a security risk? Just the involvement of

---

[55] Taylor, Elaine, "Were They Spies For Roosevent?", Air Classics Magazine. Vol. 24. No. 2, February 1988..

[56] Goerner, page 132.

such an agency is a glaring indicator of political undercurrents in the disappearance of the aviatrix.

After Devine published a book detailing his investigation, he continued his efforts to probe the Earhart disappearance, as discussed earlier in this book. Ads in various publications such as "Leatherneck", catering to military personnel, produced some startling results.

In 1975, Devine received a letter from a William Gradt of Chicago, Illinois. Gradt had read Fred Goerner's book and took exception to General Greene's statement to Goerner that he knew nothing of Earhart or her plane. According to Gradt, he had been given access to a Marine Corps archive somewhere in Virginia, where during his visit there, he spent eight hours going through Earhart files. Even more surprisingly, he stated that someone inadvertently released copies of extremely sensitive documents regarding Earhart to him. Gradt did not indicate why he was at the archive or in what capacity, but his inference was that what he had seen in the Earhart files, and in the documents released to him, proved that General Greene's remark to Goerner was not true.

Gradt also noted that he did not realize for years the significance of the documents until he showed them to a friend, who was able to relate them to other materials and make sense of them.

According to Devine, Gradt did not write again until January 1979, when he said, in part:

"Another reason I make this request is that I have for years held copies of official documents with names that tie in directly to your story. These are most critical due to the nature as the names indicated are of the top three levels of government at the time. They completely contradict and prove beyond any doubt that the information released by them so far is a complete hoax.

"These are now under lock and key and will be used only as a last resort after all is out in the open. No one to date has ever come close to making mention of these, the information they contain or persons involved. The mere exposure would discredit many beyond belief and this is true fact. This is why I do not seek publicity or financial gain and sensationalism I can do without. I only want to see the entire matter cleared up in my lifetime and yours."

Over the next year, Gradt wrote several other letters to Devine indicating that he had been attempting to get further information about Earhart from an unnamed government agency. He also indicated that the agency had acknowledged in a letter to him that they were aware that sensitive documents had been released to him and that they had not figured out how this had happened.

Amazingly, Gradt then quoted another portion of the letter in which the unnamed agency offered to allow Gradt to view a report of some sort regarding Earhart and Noonan's disappearance, but required him to first sign a draconian secrecy agreement. According to Gradt, there were 450 other documents on the Earhart disappearance being held by the unnamed agency.

In a subsequent letter to Devine, Gradt stressed that the documents he possessed contained information not brought out by any other researchers to date and that the information in the documents was part of a "gigantic puzzle".

He concluded that he had been unsuccessful in gaining access to any material other than what had already been released to him. And maddeningly, Gradt did not supply Devine with actual copies of the documents he said he held, although, in an earlier letter, Gradt had included a photograph of an etching found on a wall in Garapan prison in 1944 during the invasion. That etching later became the centerpiece of the artwork for the cover of Devine's book, "With Our Own Eyes".

Strange as it may seem in places, Gradt's information, which was elicited by an ad placed by Devine in a Marine Corps magazine, seems to at least help confirm that the government has been holding extensive files on the Earhart disappearance, which contain the true facts of the incident.

Another evidence of a cover-up is the series of articles appearing in the Woodbridge, New Jersey News Tribune in 1982. These articles contained many photos of a woman who was similar in appearance to Irene Bolam, but they were not Bolam. Tod Swindell's photo study determined this with forensic techniques. The article in the New Jersey newspaper raises some significant questions. If Irene Bolam was merely an elderly old-time aviatrix, and not Amelia Earhart, why would someone publish a whole series of photos of another woman and purport them to be Bolam?

Error on the part of the newspaper can fairly well be ruled out because we are dealing not with one or two photos, but with numerous photos. In fact, only a couple of photos in the articles perhaps, might have been of the woman Joe Gervais met in Long Island under the name of Irene Bolam. The only logical answer is that someone wanted very much to discredit the idea that Irene Bolam was Amelia Earhart and confuse the issues.

Appearing in 1982, some 45 years after the disappearance of Amelia Earhart, this otherwise accurate and provocative series of newspaper articles seems to be troubling if not strong evidence of a long-term cover-up.

The most recent indicator I have encountered of an irregularity is the microfilm of the records regarding the Coast Guard search for Earhart. A part of the microfilm contained copies of the ITASCA radio log, which was supposed to have been donated by the Bellarts family in 1975.

However, near the end of the preparation of this book, when I printed copies of the radio log from my copy of the national archive microfilm to use them as an appendix for this book, I immediately noticed a significant discrepancy.

They did not remotely match the copies that Dave Bellarts had previously given me of the originals which his father, Chief Bellarts, had held from 1937 until his death in 1974. Shortly afterward, in 1975, the originals were donated to the U.S. National Archives. The copies Dave had given me were typed in all caps in a "communications" typestyle and contained the signatures of all the radio watchstanders above their printed names.

When comparing the archive pages with the Bellarts pages, I could immediately tell that the archive pages were different from the Bellarts pages in two respects:

They were in a different typestyle and there were slight differences in the content.

Nowhere on them were signatures of the original watchstanders.

I let Dave Bellarts know of the discrepancies and he was extremely concerned. He immediately launched his own investigation to determine why the radio logs that his family had donated to the national archives had apparently been substituted.

As of this writing, Dave is still making inquiries. However, I felt these discrepancies should be revealed in this report. Why would something be substituted for the radio log pages donated by the Bellarts family on the official microfilm? Is this another echo of a cover-up?

One immediate reason for a cover-up may well be the government's fear of embarrassment over the loss of an internationally beloved public figure. Their preparations and back-up provisions were lame, and, to boot, their rescue took over a week to get started. The ineffectuality of the ensuing air and sea search is now legendary.

Additionally, there were even more serious concerns. In late 1936, U.S. cryptanalysts managed to break enough Japanese codes, that by the spring of 1937, there was a considerable flow of high level intelligence to the White House.⁵⁷ This was an extremely critical advantage, one that would be blown if the U.S. Government made any moves which even hinted that they could read Japanese communications.

With historical hindsight, we also now know that the Japanese were lying in wait for Earhart, having dispatched ships to watch for her overflight.

For all intents and purposes, the aviatrix was set to fly straight into a trap the minute she flew anywhere near the Japanese Mandate Islands.

Overall, when one looks at what happened near the end of the last flight, it's not surprising that the matter would have been deemed hugely sensitive for the near term. If there had in fact been an armed Japanese attack on Earhart's ship, this could have been held as an act of war. The presence of U.S. military ships in the area, participating in an intelligence-gathering operation with Earhart would have been dynamite.

Had it been reported in the media that Japanese fighters had attacked a flyer as beloved as Earhart, there would have been an enormous uproar, discreditation of the Roosevelt administration, and the danger of war.

It is clear that most of the initial cover-up was to save embarrassment for the government over the ineptitude of their preparations for Earhart's flight and their attempts to rescue her.

But after two-thirds of a century, these considerations have long since ceased to be valid. And yet these events still remain classified.

A partial answer lies in an unwritten political rule in the government: "Under no circumstances will a previous administration be embarrassed by the current one, if it can be avoided".

Beyond the above reasons, there is a now well-documented economic relationship between the Japanese and American governments. Neither the U.S. government nor the Japanese government has seemed anxious for the culpability of the Japanese government to be aired. Even at this late date, they may feel that such revelations might jeopardize international relations and be "bad for business".

Many questions remain. There is the matter of radio transmissions after Earhart's ditching, lasting until July 9th or later, which only ham radio operators owned up to hearing. There is also the incredibly bizarre case of Wilbur Rothar and Earhart's scarf. In addition, there is the mystery of the unclaimed letter in Jaluit and the enigmas of Irene Bolam and William Van Dusen. And then there is the matter of the documents which William Gradt is said to have received accidentally by the U.S. government. Or

---

⁵⁷ Kahn, David, "The Codebreakers" (NY: MacMillan Publishing, 1967, 9ᵀᴴ Ed.)

was it an accident? Did someone try to leak part of the truth and finally end the cover-up?

Near the completion of this work, I came across something that is an astute comment on cover-ups in general, and the Earhart case in particular. It also uncannily sums up the thesis of this book. Recently, two test pilots who had worked with Clarence Kelly Johnson, the onetime technical advisor of Amelia Earhart's last flight and legendary figure in Lockheed's Burbank, California "Skunk Works", were interviewed by Ron Reuther.

The pilots described Johnson as unique and in total control of whatever went on in the "Skunk Works". One of them remarked that Johnson was excellent on security, and reminisced about a remark that Johnson had once made:

"If you want to keep a secret, don't call it a secret."

Afterword

It is amazing how many lives one person can sometimes touch and affect. Nearly seventy years after her disappearance, Amelia Earhart continues to touch and affect lives. Perhaps that is one true measure of greatness.

The odds are building that eventually, the rest of the pieces of the Earhart puzzle will find the light of day. One tantalizing possibility is the Putnam papers, which may well contain the answers to this baffling political mystery. Then there are the sensitive documents Don Gradt claimed to possess.

Another possibility is the set of fingerprints that Diana Dawes, a close friend of Irene Bolam, claimed to possess. If they surface one day, those fingerprints may be a major clue in the Earhart mystery.

Moreover, I have found from my participation in the Amelia Earhart Discussion Group that inquiry into the disappearance of the late great aviatrix has not ceased by a long shot. Even as these words are being written, someone somewhere is rifling through ancient files or talking to someone. The digging continues. It is my own opinion, given this unrelenting private inquiry, that the mystery of Amelia Earhart's disappearance will one day be solved.

Then hopefully, Earhart's and Noonan's heroism will be revealed and acknowledged publicly by the government. The whole affair will be placed into historical perspective, and, hopefully, we will have the full picture of all the sleight of hand, misdirection and even treachery that took place back in the spring and early summer of 1937.

Also, when the rest of the truth comes out, we may more fully understand who George Palmer Putnam was and what his place was in the mystery. For, the dynamics between Earhart and her husband, and his place in the whole affair, are almost as interesting and indecipherable as Earhart's disappearance.

Dismissed by the press at the time as "the lens louse" for ubiquitously popping into the frame during photo shoots, Putnam was in reality a smooth, sophisticated, master promoter and manipulator who seemed to be in strong control of Earhart's career. Was he a benevolent, devoted manager or a latter day control freak, keeping Earhart and Noonan out of the information loop regarding major details of the flight?

For now, all we can state is that she may have perished on Saipan, but that it is also quite possible that Amelia Earhart survived her ordeal and returned to the United States after WWII to live under an assumed name. It appears less possible that Fred Noonan might have survived. If Amelia Earhart assumed the identity of Irene Bolam, why didn't she leave behind an account of her amazing adventure for posterity? If Fred Noonan survived, why didn't he? We are left to wonder why she and possibly Fred Noonan might have spent half their lives in anonymity.

In the truest sense of the word, the Earhart disappearance is like the hall of mirrors in an amusement park. Just when you think you have found the answer, you find yourself walking headlong into another mirror. Classic misdirection. A non-mystery created for political reasons to cover up ineptitude and preserve careers, whose answer has been known all along by those in the right circles.

Pure legerdemain.

APPENDICES

Appendix I

Timeline of Events Relating to the Disappearance of Amelia Earhart

by
Ronald T. Reuther
June 2004

[Author's note: this timeline should be regarded as a research document, and more of a work in progress than a finalized document. As incredibly informative and impressive as it is, it still contains some material which is speculative or which may change as Ron uncovers new information.]

This is an updated version of earlier messages assembling a timeline of many Earhart involved people and events. It includes material from Joe Klaas, Ron Bright, and myself, with input from a number of others. The comments about James Francis Kelley were largely extracted from his *Memoirs of Msgr. ("Doc") J.F. Kelley,* 1987 transcribed from his dictated memoirs and copyrighted by J.F. Kelley. Several major events described in Kelley's book are erroneous or confused. I understand Kelley (age 85 at the time) was rather blind at this point and showing signs of Alzheimer's disease. However, there is no doubt of his many achievements and friendships with people of influence and status. In a number of references there are discrepancies and variations of critical dates, especially in those pertaining to Jackie Cochran, but also including otherwise good biographers of Earhart's life. I have attempted to cite dates based on my best interpretation of all the references I have reviewed.

Note the geographic and date concentrations in the Earhart saga. Some may be simply coincidences, but others may not.

I welcome any comments, suggestions of errors or additional data.

1930s, i.e., in the greater New York City, Long Island, NY, and Newark, NJ Metropolitan Area: G. Putnam, A. Earhart, Odlum, Cochran, I. Craigmile, Kelley;

1931-32 in Miami, FL, Floyd Odlum, Jackie Cochran, Fred Noonan, (and possibly Alvin Heller), all there off and on especially during winter months, (Heller and Noonan worked for PAA);

Early June 1932 in Rome, G. Putnam, A. Earhart, Spellman, Pope Pius XI;

2nd week of June 1932 in Brussels area, G. Putnam, A. Earhart, Kelley, and possibly Archbishop Pacelli who later became Pope Pius XII;

July 1937 in the mid-Pacific, reports of Earhart and Noonan's disappearance and reports of their survival by island natives and others. Then Department of Interior Field Representative Richard B. Black has an interesting background. He was in charge (May 1936Aug? 1937) of administration of American Equatorial Islands of Jarvis, Baker, and Holland Islands and personal representative of George Putnam the USCG ITASCA and on Holland Island July 2, 1937, and an amateur radio operator with code speed of 30 words per minute in 1933. Black beginning with Earhart's disappearance and thereafter for the rest of his life was involved with geographic locations and specific time references that match the supposed disappearance and survival route of Earhart. He gave conflicting accounts of his knowledge about her disappearance for the rest of his life. Black entered the US Navy in 1938 and: became an Intelligence Officer at Honolulu in 1941; in September 1943 involved in reoccupation of Baker Island; in January 1944 involved in invasion of Kwajelein; in June 1944 involved in invasion of Saipan; during September 1945-January 1946 in invasion and occupation of Japan; May 1946-1948 Special Representative, Pacific Ocean Areas, US Commercial Company, involved in rehabilitation of economy of Micronesia and Marianas Islands; 1950 in Korea with 8th Army, then retreated to Japan for research; February 1953-September 1965, Office of Naval Research; November 1959 retired as Rear Admiral USNR. Black throughout his career was close associate of Rear Admiral Richard Byrd, who in turn was a close friend and supporter of Amelia Earhart and George Putnam (by 1939 Byrd and Putnam were estranged). General Douglas MacArthur, Major Eisenhower, and Captain Lucius Clay (Eisenhower and Clay later became generals), all were stationed together in Manila in the Philippines in 1937 and in position to be knowledgeable about Earhart's status (could they have had access to Japanese or American military intelligence concerning Earhart?).

December 1937 May 1938, Spellman, Putnam, Vincent Astor, Kermit Roosevelt II (a friend of Putnam's) all in Panama Canal-Galapagos Island area during the same time frame.

1944 in the Marshall Islands (February) and Saipan (June July); Admiral Nimitz, USMC General Vandegrift, USMC General Tommy Watson, USMC General Erskine Graves, USA later General Lucius Clay, many other US military officers and enlisted people involved in reports of Earhart in the hands of the Japanese as the US military invaded the Marshall Islands and Saipan.

About August 18-26, 1945 Cochran, Spellman, Nimitz (then ranking Navy officer in Pacific), General Barney Giles (then ranking USAAC officer in the Pacific as deputy to General Arnold), General "Tooey" Spaatz, General LeMay, General Doolittle, and other flag officers are all on Guam, and at least most of them are in contact with each other including Cochran.

August 26 - September 1945 activities in the Far East, Cochran, Spellman, McCrary, Catholic Church, General MacArthur, Admiral Nimitz, Admiral Byrd, and Commander Richard Black all in Japan at or shortly before and after peace signing. Spellman was well connected to all US Presidents from FDR to Truman, Eisenhower, etc. until his death. Cochran during this time was on assignment by General "Hap" Arnold, Chief of Army Air Forces, and a Liberty Magazine war correspondent. A. Earhart had written for the magazine in February 1937, before her disappearance, and George Putnam wrote some articles for the magazine for a couple of years after her disappearance. In 1948 Atlas Corporation (Floyd Odlum and Jackie Cochran) bought *Liberty Magazine*.

1952 and afterwards in greater NYC area, Jackie Cochran and Floyd Odlum? (splitting time between NYC and Indio, CA), Tex McCrary, Cardinal Spellman (living in St. Patrick's Cathedral?), Msgr. Kelley, Irene Craigmile Bolam, General Douglas MacArthur (living in Waldorf Astoria Hotel) and was a candidate in the Presidential elections, General "Ike" Eisenhower in touch with McCrary and Cochran and campaigning for Presidency in and out of NYC, General Lucius Clay in area. MacArthur, Eisenhower, and Clay had all been close Army associates thruout most of their careers. Others such as Admiral Black, senior USMC generals, Col. Balchen, etc. were in the Washington, DC area. Admiral Byrd was in Boston.

August 78, 1965 at Westhampton, Long Island, NY and succeeding events concerning Joe Gervais, Irene Bolam, Msgr. Kelley, and Earhart;

1970-71 concerning Klaas/Gervais book, Irene Bolam, and Earhart;

1982 concerning Irene Bolam and Earhart.

Ron Reuther

1887 September 7, 1887 George Palmer Putnam born in Rye, NY.

1888 October 5, 1888 Richard Evelyn Byrd born, Winchester, VA.

1889
May 4, 1889 Francis Joseph Spellman (later Cardinal Spellman) born in Whitman, Mass. Graduated Fordham University in NYC (1911) and the American College at Rome and ordained a Roman Catholic priest May 14,1916. He was made a Monsignor in 1926, and a Cardinal in 1946.

1892 Floyd Odlum born in Union City, Michigan.

1893
April 4, 1893 Fred Noonan born in Chicago, Cook County, IL.

1896
Irene Rutherford O'Crowley born in 1896 (became lawyer). Had offices in Newark, NJ for most of her professional life. Also had office in NYC. Was aunt of Irene Craigmile Heller Bolam.

1897 July 24, 1897 Amelia Mary Earhart born in Atchison, KS.

1899
October 23, 1899, Bernt Balchen born Tveit, Topdal, Norway. Pilot on "America," Commander Byrd's transatlantic flight, NY-Paris, 1927; chief pilot of Byrd's 1st Antarctic Expedition 19281930, aviation friend and advisor to Earhart 1932+, transferred from Royal Norwegian to Royal Air Force to USAAC in 1941. Commanding Officer of Air Operations Against German Forces in Scandinavia (OSS) 1941-1945.

1900
December 29, 1900 Muriel Earhart (later Muriel Earhart Morrissey) born, after Amelia, in Atchison, KS.

1902
July 27, 1902, James Francis Kelley born in Harrison (Kearny), NJ.

August 10, 1902, Richard Blackburn Black born at Grand Forks, North Dakota.

1904
October 1, 1904 Irene Madalaine O'Crowley born at Newark, NJ at a residence, no birth certificate on file. Msgr. Kelley was intimately associated with the Newark Diocese for almost all of his adult life.

1905 December 23, 1905 Howard Hughes born in Houston, TX.

1906
May 11, 1906 Jackie Cochran (Bessie Lee Pittman) born in Muscogee, FL.

1908
Fall 1908 George Putnam attended UC Berkeley, CA.

1909
Early 1909 George Putnam met Dorothy Binney on John Muir Trail hike near Mt. Whitney, CA., he settled in Bend, OR.

1911
July 14, 1911 Frank (Francis) Spellman graduated from Fordham College, NYC. Departed for college in Rome shortly afterwards.

October 26, 1911 Putnam married Dorothy Binney in Sound Beach, CT. Within months they returned to live in Bend, OR.

1916
January 14, 1916 Fred Noonan as able bodied seaman departed Boston on American ship arriving Buenos Aires July 6, 1916.
May 14, 1916 Francis Joseph Spellman graduated from the North American College in Rome and ordained a priest. He returned to Boston

and affiliated with the Cathedral of the Holy Cross. He had conflicts with the Boston Cardinal William O'Connell.

June 1916 George Putnam is called to active duty from Oregon with the Oregon National Guard in the Mexican Border War.

July 12, 1916 Fred Noonan, as 3rd mate, departed NYC on British steamer *Carracas* arriving December 9, 1916 Valparaiso, Chile.

Late December 1916 Fred Noonan departed NYC on American ship *SS New York* from NYC to Liverpool.

1917
January 14, 1917 Fred Noonan arrived at NYC aboard ship *SS New York*.

April 1917, U.S. entered WWI.

June 1917, Putnam was assigned as a special agent with the Department of Justice in Washington, DC for 18 months.

Fall 1917, Earhart at O'Gontz School (a finishing school) in Rydal, PA a North Philadelphia suburb. Met teacher(?), Elizabeth Vining (age 14-16), who resigned? about that time, and in 1946 went to Japan where she became an English tutor to the children of Emperor Hirohito till 1950.

1918
November 1, 1918 James Craigmile (Irene Craigmile Heller Bolam's first husband) completed School of Fire for Field Artillery at Ft. Sill, Oklahoma.

November 11, 1918 Armistice signed, ending WWI.

December 1918 Putnam commissioned a 2nd Lt. after attending Field Artillery Officer Training School at Camp Taylor, KY.

December 1918 Putnam became President (Vice President?) of G.P. Putnam's Sons, Publishers in NYC.

Beginning in February Amelia Earhart lives briefly in Northhampton, Massachusetts (near Boston) with her sister Muriel.

1919
Fall 1919, Amelia enrolls in premed course at Columbia University, NYC; makes friends with fellow student Louise de Schwinitz; (per Goldstein and Dillon).

Lt. Commander Isoroko Yamamoto arrives at Harvard University, Boston, to study English and economics. Finishes in 1921. Returns to US as a Captain and Naval Attaché in Washington, DC in 1925-1928.

1920
Summer 1920 Amelia joins family in Los Angeles, Southern California.

1921
May 16, 1921 (per Lovell?), Dec 15, 1921? (per Rich), May 19,1923 (per Goldstein and Dillon) Amelia Earhart receives NAA pilot's license in Los Angeles.

Francis Spellman became Chancellor of the Boston Diocese. **May 15, 1923** Earhart received FAI pilots license in Los Angeles (Rich), May 19,1923 (Goldstein and Dillon).

1924
Spring (Lovell?), Amelia returns with mother and sister to live briefly in Medford, Boston area.

June 1924
Kelley graduated with A.B. from Seton Hall College, S. Orange, NJ, goes to Europe.
Francis Joseph Spellman made archivist of Archdiocese of Boston.

September 24, 1924 (Rich) Earhart starts 2nd year at Columbia University, NYC.

There is a strong possibility that Earhart and Irene O'Crowley [Craigmile] were at Columbia University, NY in 1924 at the same time and may have met there.

1925

Summer, Amelia attends Harvard Summer School in Boston; September began teaching English to foreign students in University of Massachusetts extension program (Long and Long).

November 2, 1925 Francis J. Spellman made American Assistant to the Papal Secretariat of State (the Vatican) and departs Boston, in Rome till 1931, returned to U.S. in 1932 after some months in Paris and Belgium?.

1926

Summer, Amelia began teaching to foreign-born pupils at Denison House, Boston. Amelia's close friend, Louise de Schwinitz, from Columbia University was interning at a Boston hospital. Amelia remains in Boston until April 1928.

Francis Spellman made a Monsignor in Rome.

Winter 1926, Richard B. Black visited Japan as part of "Floating University" cruise to 47 ports in 23 countries.

1927

Early 1927 for some months A. Earhart taught foreign students (including? then Lt. Cmdr. Isoroko Yamamoto, IJN) in the extension program of the University of Massachusetts (Goldstein and Dillon; *Soaring Wings* by Putnam).

June - July 1927 Commander R.E. Byrd made NY Paris flight with Bernt Balchen , Bert Acosta, and George Noville. AE. and GPP are friends with Byrd and Balchen. *Skyward* by Byrd previously published by GPP in 1928.

Irene O'Crowley marries James Charles Craigmile in Pompton Plains, OH, in 1927. James Craigmile dies September 23, 1931 in Detroit. Irene returns to NYC area.

1928

Admiral Richard Byrd (friend and commander of Bernt Balchen) who then lived in Boston sold the *Friendship* to Mrs. Frederick Guest who sponsored the flight in this aircraft which carried Earhart across the Atlantic in 1928.

Putnam identifies Earhart as potential female first air passenger to fly across the Atlantic with Stultz and Gordon in the *Friendship*.

June 19, 1928 Earhart makes flight as passenger in the *Friendship* across Atlantic, landing at Burry Port, S. Wales; much publicity.

Late June 1928 Earhart and crew receive official "ticker tape" parade and reception in NYC.

Early July 1928 Earhart and crew receive large official receptions in Boston; Medford, Mass; and Chicago.

July 8, 1928 Kelley ordained a priest at American College, Louvain, Belgium, Remained in Europe until 1935. Met Archbishop Pacelli in Belgium and taught him English. Pacelli was the Apostolic Delegate for Belgium in the early 1930s. He later became Pope Pius XII.

1929 Cochran began living in NYC.

Wall Street collapse; depression continues through 1930s.

Fred Noonan working for NYRBA Airlines in .Miami, FL

December 19, 1929 George Putnam divorced by first wife Dorothy Binney; marriage had produced two children.

1930
Fred Noonan working for PAA in and out of Miami till March 1935.

A. Earhart at Cleveland Air Races. voted first President of 99s.

November 21, 1930 General Douglas MacArthur became Chief of Staff, US Army in Washington, DC through October 1, 1935.

1931
February 7, 1931 Putnam (2nd marriage) marries Amelia Earhart in George's mother's home in Noank, CT. [License issued in Noank, CT.] George and Amelia immediately begin living in his Rye, NY home, but begin sharing time between NYC area and southern CA. Earhart working at PAA headquarters in NYC.

Beginning in 1931 George Putnam worked 4 years in Hollywood and New York City as head of Paramount Studios Editorial Board and continued work for Paramount thru 1938.

December 1931 Jackie Cochran and Floyd Odlum, President of the Atlas Corporation and among the 10th wealthiest men in the U.S, meet in Miami, FL , both live in NYC. They begin sharing time between NYC and southern CA with some winter time in Miami Beach. They marry (Odlum's 2nd marriage) May 11, 1936 secretly in Kingman, AZ.

1932
May 20-21, 1932 Earhart flies Atlantic solo in Lockheed Vega, landing at Culmore, N. Ireland, much publicity. Bert Balchen assisted her in preparing the airplane and flying it to Nova Scotia. AE gave high praise to Admiral Byrd for his assistance.

June 1932 Putnam and Earhart have audience with Pope Pius XI in Rome followed by visit with King and Queen of Belgium in Brussels in June, 1932. Msgr. Kelley is studying at Louvain, Belgium at the time, and perhaps Archbishop Pacelli (later Pope Pius XII) was there at the same time.

July 30, 1932 Spellman was made an auxiliary bishop of Boston and Pastor of Sacred Heart Parish, Newton Center, Massachusetts until April 15, 1939 when he was made the archbishop of NYC by Pope Pius XII.

Cochran and Irene Craigmile working as beauticians? with cosmetics in New York City in 1932. [as an aside note that Miss Genevieve Crowley [could this be a mistake in spelling by Cochran?], who was office manager of Antoine's in NYC came to work for Jackie Cochran in 1929? and stayed until she retired at the end of 1953].

Earhart, Irene Craigmile, Jackie Cochran, and Viola Gentry all flying in the same time period, 19321933, at Roosevelt Field, NY and maybe Floyd Bennett Field, NY. Alvin Heller and Viola Gentry's husband are Craigmile's instructors at Roosevelt Field. Not yet clear if Cochran knew any of the above women at that time, but likely.

August of 1932 Cochran first soloed and received pilots license August 17, 1932 at Roosevelt Field, Long Island, NY. Cochran did meet George Putnam (then married to Amelia) in 1932 soon after (Cochran is quoted in her biography as saying she did not meet Amelia until 1935).

August 21, 1932 Cleveland Air Races, Irene Craigmile there? per Ann Pellegreno. Amelia Earhart there?

In 1932 Odlum is largest financial backer of FDR in his first presidential campaign and must have known FDR personally then (they both lived and worked in the NY area). Odlum later after backing Truman, switches to Republican Party.

Late September 1932 Francis J. Spellman returned to U.S aboard the liner *Rex* from assignment at Vatican in Rome via several months stay in Paris and Belgium?. Spellman over the next several years accumulated the largest collection of Amelia Earhart stamps in the world, which is now housed in the Spellman Philatelic Museum and Library in Boston.

October 1932 Earhart voted Outstanding American Woman of the Year.

Irene Rutherford O'Crowley handled legal affairs for Earhart in the 1930's. Claims she first met Earhart in Europe when she and niece [Irene O'Crowley, later Craigmile, Heller, Craigmile, Bolam] were traveling in Europe and Earhart had just flown solo across the Atlantic to Europe. Irene Rutherford O'Crowley claims her niece did not meet Earhart then or later.

Irene Rutherford O'Crowley's *Pilot Publishing Company* associates are Miss "Craig" and Miss "Mile," which together spell "Craigmile," the first of Irene O'Crowley Bolam's three married names.

1933
February 1933, Major Dwight Eisenhower becomes Chief Military Aide to US Army Chief of Staff, Douglas MacArthur in Washington, DC, until September 1935, when he accompanies MacArthur to the Philippines as his senior aide and as Assistant Military Advisor to the Philippine Commonwealth.

May 27, 1933 Irene Craigmile (Heller Bolam) received her pilots license at Floyd Bennett Field, Long Island.

July 2, 1933 A. Earhart at Cleveland (National) Air Races (Morrissey and Osborne).

August 6, 1933 Amelia, Gene Vidal, Paul Collins, and Sam Soloman start National Airways (Boston & Maine Airways), which became Northeast

Airlines. In 1938 Atlas Corporation (Floyd Odlum, Jackie Cochran's husband, was President), began buying shares of Northeast obtaining a majority share in the 1940s, and in 1944 Jackie Cochran became a member of the Board of Directors.

August 15, 1933 Alvin Heller and Irene Craigmile elope (to Columbus, OH), annulled in 1940. Irene reassumed her name as Irene Craigmile and returned to NY, NJ area. Alvin Heller worked at some time for Pan American Airways in Miami, he was also President of the Greater Miami Aviation Club in 1934? (per Tod Swindell).

Floyd Odlum helped reorganize Paramount Studios and bought major quantity of debentures of Paramount. Odlum sold Paramount and bought RKO Studios in 1935.

1934

January 11-13, 1934 Howard Hughes placed first in All American Air Races in Miami.

March 5, 1934 Clarence Heller born in Newark, NJ. Supposedly son of Irene Craigmile Heller (Bolam).

April (Antarctic winter) 1934 Richard B. Black met Admiral Byrd for first time in Antarctica as radio operator, assistant scientist and worked for him thru 1935 and again in 1939 March 1941, and again in November 1954 October 1957.

July 21, 1934 Kelley received Degree of Licentiate in Philosophy at Louvain, Belgium. Did graduate work there in Psychology and Psychiatry.

March 27, 1935 Fred Noonan arrived at Alameda, CA as navigator for PAA introducing PAA Pacific service.

July 8, 1935 Kelley received Ph.D. in Philosophy and Psychology, returned to U.S. in August 1935 as head of Department of Philosophy at Seton Hall College and Professor of Psychology at Immaculate Conception Seminary, Darlington, NJ.

During the early and mid thirties Earhart often flew in and out of Newark Airport and set a number of flight records in doing so.

Cochran is quoted as saying she "first met Earhart in 1935."

August 30 - September 2, 1935 Earhart competes in National Air Races in Cleveland.

September 18, 1935 Howard Hughes set world speed record in Santa Ana, CA in H1 *Racer.* Earhart and Paul Mantz flew escort for official timing in 2 separate airplanes. Sometime in the month Hughes leases Northrop Gamma from Jackie Cochran.

September 1935 Major Dwight Eisenhower accompanies General Douglas MacArthur to the Philippines as his senior aide and remains there until December 1939.

November 22, 1935, two Japanese nationals secretly slipped aboard the PAA *China Clipper* at Alameda, CA. They were trying to miscalibrate the direction finder, when FBI agents apprehended both of them.

1935
January 5, 1936 PAA *China Clipper's* hull was slashed open by submerged iron rods embedded in concrete pedestals found in the taxi lane at Alameda, CA. Likely sabotage by Japanese.

January 13, 1936 Howard Hughes sets transcontinental speed record Burbank to Newark in Northrop Gamma leased from Jackie Cochran.

May 11, 1936 Cochran and Odlum secretly marry in Kingman, AZ. Kingman, AZ was stop point for Cochran in Bendix Trophy Race in year?. Also Kingman was transmitting station picked up by Paul Mantz in radio check out when airborne over Hawaii in 1935 from Earhart's Lockheed Vega just before her departure solo for Oakland. I believe Earhart also was able to talk with same Kingman station while she was enroute to Oakland in the Vega.

Earhart and Putnam are good friends with Franklin and Eleanor Roosevelt (Putnam was a boyhood acquaintance of FDR and beginning in 1936 FDR arranged strong US Government support for Earhart's last flight, 1937). Beginning in 1936 (possibly earlier) Cochran may have been friends with FDR. Odlum had been FDR's biggest financial backer in FDR's first campaign for President in 1932. Odlum and Cochran began funding some of Earhart flight costs including the last flight. Earhart and Putnam spend time at Indio Ranch with Odlum and Cochran as does Howard Hughes, but it is unknown if Hughes is ever present when Putnam's are there.

July 1936 Kelley appointed President of Seton Hall College (271 students in Liberal Arts and Sciences), the youngest President ever appointed in U.S. to an accredited college. Kelley became affiliated with Mt. St. Dominic Academy, Caldwell, NJ in 1936, (same school that Irene Madelaine O'Crowley (IB) had attended in the 7th grade in 1916).

Monsignor Francis Spellman becomes good friend of FDR as result of visit of Cardinal Pacelli (future Pope XII and good friend of Spellman's and Msgr. Kelley) to US and with FDR.

December 1936 Fred Noonan leaves Pan American Airways after having been their chief navigator since 1935. He had made 18 transpacific flights with PAA.

1937
US Army officer (Captain?) Lucius Clay (assistant to Eisenhower), and Major Dwight Eisenhower, both later became generals, are on staff of General Douglas MacArthur in the Philippines thru all of 1937 (could they have had access to US and/or Japanese naval intelligence re Earhart?). Later they are all involved in various events in Washington, DC in the military and NYC in civilian life, and thus possibly knowledgeable about disappearance and fate of Earhart and Noonan. Clay had been in Washington, DC as head of Army public works since 1933 under guidance of FDR.

January 31, 1937 Fred Noonan hired as assistant navigator to Harry Manning on Earhart crew per Loomis and Ethell, p. 50.

March 1937 W.T. Miller of the Bureau of Air Commerce, working under Eugene Vidal, a close friend of Earhart's and Director of the Bureau, locates in Oakland, mans an office in downtown Oakland and helps plan and coordinate Earhart flight. He had been initial officer in colonizing the American Equatorial Line Islands (Jarvis, Baker, and Howland islands) in 1935, and then in 1936, assisted and was replaced by Richard Black as administrator.

March 17, 1937 Earhart, pilot; Paul Mantz, copilot; Harry Manning, radio operator and navigator; and Fred Noonan, navigator, depart Oakland

(1st attempt RTW) in Lockheed Electra 10E and on March 18 arrive at Wheeler Field, Hawaii.

March 20, 1937 Amelia Earhart, Fred Noonan, and Harry Manning attempt takeoff at Luke Field, Ford Island, Pearl Harbor, Hawaii, (1st attempt RTW, westbound). The aircraft groundloops and is badly damaged.

Several people (PAA pilot Mark Walker, radio operators McMenamy and Pierson, Earhart friend Elmer Dimity, and researchers and authors Art Kennedy and Robert Myers), are quoted as saying Earhart told them that she was making the flight at the request of and under the control of high government officials.

Earhart spends time with Jackie Cochran and Floyd Odlum at their ranch in Indio, CA. Howard Hughes is also a frequent visitor to the ranch, though it is unknown if he was there when Earhart was there.

April 15 thru May 2, 1937 Earhart in NYC area.

General Oscar Westover and Bernard Baruch meet with Earhart at March AFB, Riverside, CA on two, possibly more occasions (per DeCarie to Gervais). Bernard Baruch, advisor to FDR, may have met with Earhart in her N. Hollywood home (per DeCarie per Gervais) after her return from the Luke Field ground loop and the 1st attempt departure. Baruch contributed some funds for 2nd attempt. Baruch was also a close friend of Monsignor Spellman and FDR. During this time, W.T. Miller of the Bureau of Air Commerce moves into Earhart home in N. Hollywood and is again strongly involved in planning the 2nd attempt RTW flight.

May 20, 1937 Amelia Earhart departs Oakland CA (Noonan joins flight in Burbank next day) (2nd attempt RTW eastbound via Burbank, Tucson, New Orleans, and Miami), in opposite direction to the first attempt.

July 2, 1937 Earhart and Noonan disappear in mid-Pacific in vicinity of Howland Island? George Putnam is devastated and makes much effort to find out what happened and arranges some searches by the US Navy in the Phoenix Islands and others near the Gilbert Islands.

Then Department of Interior Field Representative Richard B. Black in charge (May 1936Aug? 1937) of administration of American Equatorial Islands of Jarvis, Baker, and Howland Islands and personal representative of George Putnam on the USCG *ITASCA* and on Howland Island Jul 2,

1937, and an amateur radio operator with code speed of 30 words per minute in 1933; entered the US Navy in 1938 and became an Intelligence officer at Honolulu in 1941; in September 1943 involved in reoccupation of Baker Island; in January 1944 involved in invasion of Kwajelein; in June 1944 involved in invasion of Saipan; during September 1945-January 1946 in invasion and occupation of Japan; May 1946-1948 Special Representative, Pacific Ocean Areas, US Commercial Company, involved in rehabilitation of economy of Micronesia and Marianas Islands; 1950 in Korea with 8th Army and then to Japan; February 1953September 1965, Office of Naval Research; 1965 retired as Rear Admiral USNR. Black throughout his career was close associate of Rear Admiral Richard Byrd, who in turn was a close friend and supporter of Amelia Earhart and George Putnam until Putnam and Byrd were estranged).

July 2, 1937 the Japanese survey ship *KOSHU* was at Ponape and on July 9, 1937 proceeded to Jaluit in the Marshall's arriving July 13, 1937. It may have proceeded to Mili Atoll to pickup Earhart and Noonan and returned to Jaluit about July 18, 1937.

July 3, 1937 Japanese Navy (per Admiral Isoroku Yamamoto) refuse to allow American vessels into mandated island areas to search for Earhart after formal request by US State Department. Japanese announce they will search for the missing flyers.

Walter McMenamy, a ham radio operator and Karl Pierson, chief engineer of the Patterson Radio Corporation, both of Los Angeles, say they were sworn to secrecy by the FBI regarding any radio messages they might receive from Earhart during the flight, and that all such messages would be transmitted only thru a government controlled radio facility on Beacon Hill in Los Angeles. Within hours of her disappearance McMenamy and Pierson were asked by the government to assist in listening and claim to hear Earhart's transmissions.

A number of radio transmissions were received by various listeners in the Pacific and in the US beginning within hours of Earhart's disappearance. Notable among these are the receptions by professional PAA radio operators in the Pacific, heavily involved radio operators Walter McMenamy and Karl Pierson in Los Angeles, radio operators on the island of Nauru, and the Royal New Zealand Navy cruiser *Achilles* , located south of the Phoenix Islands.

July 7, 1937 Marco Polo Bridge massacre by Japanese against Chinese near Peking, China. Japanese officially announce war against China as a result on July 25, 1937.

August 1949, CIA G2 document on Earhart. U.S. government had pressed Japanese for as much information as they could obtain. American intelligence agents were unable to find any Japanese Navy records pertaining to Earhart. But according to Japanese navy personnel who had supposedly searched for the Electra, the Japanese Navy's 12th Squadron, assigned to the Marshalls in 1937, was instructed by Tokyo, after a request from the U.S. government, to send the *KAMOI*, a seaplane tender, and several large flying boats, using the sea to the south of Jaluit as a central search point. Later the survey ship *KOSHU* was ordered into the area. The Japanese testified that the *KAMOI* led the rescue effort, but no traces of Earhart were found. The investigation was closed. However research by Loomis and others found that the *KAMOI* was docked in Saipan on July 2 and left July 4 for Ise Bay, Japan where it docked on July 10. Clearly the Japanese government had lied to the U.S. in 1937 and again in 1949.

July 13, 1937 report by a London international news dispatch to the [*Japan?*] *Advertiser* [in Tokyo] that a Japanese fishing boat had picked up Earhart and Noonan [in the Marshall Islands area].

July 13, 1937 Japanese survey ship *KOSHU* arrives at Jaluit, Marshall Islands.

July 1937 French explorer Eric de Bisschop describes seeing Earhart and Noonan as Japanese prisoners in Jaluit while he was held prisoner there by the Japanese. He later is released and continues on a sailing journey to Hawaii. He eventually composes and releases a "bottled message" [describing the above experience] off the coast of France which is found [October 30, 1938] and turned over to French police officials and eventually to an American naval, attaché Hillenkoetter [who became the first director of the CIA], and then to the US State Department. de Bisschop knew American naval intelligence officers in Hawaii pre WWII and later returned to Honolulu as the French Consul in the early 1940s.

July many Marshallese natives and later Saipanese natives report seeing a Caucasian male and female (flyers) in the hands of the Japanese about this time. Most reports match the appearance and clothing styles of Earhart and Noonan. These reports were made to American military personnel who invaded the Marshall Islands beginning in January 1944, and Saipan in June 1944. "It wasn't a secret among the Marshallese at the time. Most everyone knew Earhart had been captured by the Japanese. But they dared not speak of it. At least, not until after Japan's war time oppression had stopped." [Robert Reimers]

July 31, 1937 George Putnam in letter to FDR aide Marvin McIntyre says "Is there any way of ascertaining what the Japanese are actually doing, especially as regards a real search of the eastern fringe of the Marshall Islands? This is one of the most fruitful possible locations for wreckage."

Louis Ream, deputy to General William Donovan, head of OSS in WWII; and later, associated with and close friend of Allen Dulles, head of the CIA, in a letter to Goerner from Ream's nephew John Ream, was quoted as saying "It was well known within high ranking intelligence circles that Miss Earhart, at the time of her disappearance, was involved in an intelligence-gathering operation. The mission was not specifically for the US Navy, but rather was ordered at the request of the highest echelons in government. There were some serious blunders by the Navy in their attempt to provide Miss Earhart with proper guidance, and the Navy was and is determined to conceal their participation and failure in their part of the operation."

Quite some time later (a matter of months after Earhart's disappearance) some island natives arrived at Nauru Island in an outrigger canoe. They were from the Marshall and Gilbert Islands and they had a handwritten scribbled note signed supposedly by Amelia Earhart. The note stated that she had gone down and been captured by the Japanese in the Marshalls and Gilberts and she was hoping her note could be smuggled out by friendly natives. She and her navigator, Newman (sic), were held prisoners. The news was immediately reported to London, but meanwhile the local newspaper picked up the story and immediately went to press. The paper was distributed throughout the protectorate islands. Within a few days a message arrived from London classifying the story TOP SECRET. That set off a frantic search for all the papers that had been printed and sent to all the islands. The security clamp was never rescinded and no further information on the subject ever reached Nauru despite requests for information. [per Lt. Col. Jack Ralph (USAF Ret), letter dated 1989?].

Amelia Earhart Foundation established in Oakland, CA as a nonprofit foundation "to conduct an expedition and clear up the mystery and to inspire the study of aeronautical navigation and the sciences akin thereto." Elmer H.

Dimity of Oakland, a friend of Earhart's was the leader of the Foundation. According to Dimity the expedition was cancelled due to WWII.

October 7, 10 p.m., 1937 an unclaimed letter date stamped Los Angeles, California addressed to Miss Amelia Earhart (Putnam), Marshall Islands (Japanese), Ratak Group, Maloelap Island (10), South Pacific Ocean was found in the Jaluit Post Office in the Marshall Islands. A March 17, 1938 letter from Mr. Carle Heine of the Marshall Islands was published in the *Pacific Islands Monthly Magazine* dated May 25, 1938 describing the unclaimed letter to Miss Amelia Earhart (Putnam) with a return address to the Hollywood Roosevelt Hotel, Hollywood, California. Margo DeCarie, Earhart's personal assistant had been living in the Hollywood Roosevelt Hotel during September October 1937.

October 16, 1937 *Smith's Weekly*, published in Australia said "USA does Australia a Secret Service, Amelia Earhart Search Made the Opportunity, Plane Observers Over Japanese Pacific Bases, Tip Was Given Our Defence Dept." "American [naval] planes did more that just search for Amelia Earhart. They cut a wide swathe over the Pacific and circled near the Caroline and the Marshall Islands."

November 1937 George Putnam edited and published Amelia Earhart's last book, *Last Flight* with no definitive conclusions about her loss.

December 12, 1937 Japanese attack and sink *USS Panay* in river near Shanghai. Several casualties. *Panay* was an intelligence gathering ship as well as a US Navy river boat.

In *The Wild Blue Yonder, Sons of the Prophet Carry On* by Emile Gauvreau published in 1944, Gauvreau says *The Wild Blue Yonder* was reviewed by Hap Arnold and some other notables in aviation and Gauvreau certainly had substantial contacts in the field. Gauvreau was Director in Chief of a Congressional Investigation of our aircraft status launched by the Committee on Patents of the House of Representatives. This included investigation of patents, patent contracts, patent pooling, cross licensing and every phase relating to patent agreements, especially concerning the aircraft industry. He had worked with and written a book about General Billy Mitchell [died in 1936]. Gauvreau says in part: "[General "Billy"] Mitchell's warning of the Japanese menace was another reason why he was to be harassed to death because his outbursts were described as "bad for business." He was accused of displeasing our "best customer" [Japan]."

"Although they [Mitchell's remarks] would have seemed incredible at the time and were not officially discussed before the public because of Washington's fear of disturbing international relations, they may be more easily accepted now that Ambassador Joseph C. Grew has divulged the diabolical Japanese plot [prior to January 1941] to kidnap President Roosevelt on a peace conference at sea while Japan's torpedo planes stabbed at the heart of American sea power in the Pacific."

"The first omen of what was to happen in the Far East is believed in high-ranking circles of the Navy today to have come from Amelia Earhart's last call for help before her plane disappeared in July of 1937 in the vicinity of Japanese mandated territory in the Southwest Pacific. Although Washington authorities were careful to point out that her flight around the equatorial zone of the globe had "not been officially encouraged" she had been given by Government departments all technical advice available months before her last adventure. She left the United States in an $80,000 flying laboratory capable of carrying sufficient gasoline for a nonstop flight of 4,000 miles, and about 1,500 more miles than the distance from Lae in New Guinea to Howland Island, which she was reported unable to negotiate."

"For reasons of diplomacy every effort was made by the Government to keep from the public the suspicion that Miss Earhart had been shot down and all evidences of her plane destroyed after she had lost her way and flown over mandated islands which the Japanese were secretly fortifying in their vast preparations for the conquest of the Pacific. Among those convinced of this explanation of the mystery was Claude A. Swanson, Secretary of the Navy, who told Congressman Sirovich [Chairman of the Committee on Patents at the time] it was incredible under the circumstances to believe the Earhart plane could have disappeared without a trace unless every matchstick of it had been deliberately destroyed.(3).

"This is a powder keg" he said, "and any public discussion of it will furnish the torch for the explosion. I firmly believe that Miss Earhart, in trying to reach Howland Island, a speck on the map, lost her directions perhaps by a sudden shift in the wind and was brought down over territory she was not supposed to see. We are aware that something is going on there. I am not the only one in this department who feels that she saw activities which she could not have described later and remained alive.

"That is the only explanation I can reach for the blotting out of their plane and every solitary piece of her equipment. Otherwise something would have remained. The attention which Japanese newspapers paid to the smallest details of this sad flight from every mile of her progress. To speculate about this publicly probably the time it began shows the peculiar interest with which Japan followed would sever our diplomatic relations with Japan and lead to something worse."

"*(3.) Congressman Sirovich had visited Secretary Swanson to inquire about a report in Congress that Miss Earhart had been sent by the Navy on a secret mission which involved flying over Japanese mandated territory to see if it was being fortified. This report was emphatically denied.*"

"General [Frank] Andrews [then Commander General Headquarters Air Force] was among those who believed that the aviatrix had been shot down by the Japanese. "The President," he said, "would not have ordered a fighting Naval armada to the scene if the suspicion had not been officially entertained. When we start shedding American blood out there we will find out what all this means."

"The official tension reflected in Secretary Swanson's private remarks were to be more clearly defined six months later, in 1938, when *Smith's Weekly* of Sydney, Australia, asserted that the American Navy had not only taken advantage of its search to spy on Japanese mandated islands but had transmitted its information to the Australian Government. An inquiry by the Senate, bordering on the sensational, was promptly checked when, with the President's approval, Admiral William D. Leahy, as Acting Secretary of the Navy, announced that no vessels or planes of the Navy had approached the vicinity of any islands under Japanese sovereignty or mandate and that no findings of any kind regarding Japanese domain or activities were submitted in connection with the search. This may explain why nothing was ever seen of the Earhart plane. Cordell Hull, Secretary of State, issued a similar statement. Soon after Pearl Harbor the Navy was officially to approve a motion picture script which represented the execution of Amelia Earhart by the Japanese on a fortified mandated island. Although naval authorities were willing to have the film prepared and offered their cooperation, other matters interfered with its production."

"Any story speculating upon the possibility that Amelia Earhart was being held a prisoner or had been executed by the Japanese in 1937 would have been shouted down as the incoherent hallucinations of a demented fanatic. Japan's well paid publicists, a number of them American writers now in prison, were scowling into oblivion the slightest suggestion that the Japanese had aggressive aims upon the United States."

Early Dec 1937, George Putnam started his photo expedition from Los Angeles? to the Galapagos Islands, accompanied by Josephine "Joe" Berger (later Greer), his secretary. Josephine's husband, "Win" also went on the expedition. They cruised on the yawl *Athene*.

During the final weeks of the cruise, Putnam had an affair with a young female starlet, Ione Reed, who was on the expedition. The affair lasted a few more months then they broke up. The expedition returned in the Spring of 1938, probably early March. (See Backus, *Letters From Amelia*, p. 219220 and Lovell, *The Sound of Wings*, p. 303304). Some have wondered if George got aboard the *Nourmahal* (owned by Vincent Astor) on an espionage voyage late February 1938 to May 21, 1938 to the Marshall Islands at the request of FDR.

January 11, 1938 Pan American Airways *Samoan Clipper* exploded off Pago Pago, American Samoa. Possible sabotage per PAA per Ronald Jackson per Charles N. Hill.

1938
January 13, 1938 Vincent Astor cables FDR from his yacht the *Nourmahal* in the Pacific.

January 1938 Monsignor Spellman of Boston departs for recuperation cruise around South America via Peru, Bolivia, Ecuador, Chili, Argentina, Brazil, etc., returning to Boston in early April [Gannon]. [it would be interesting to know on what ship, when they traversed the Panama Canal, any contact with George Putnam, Vincent Astor and or associates, and/or *Nourmahal*? or yawl *Athene*].

Secretary of the Navy Claude A. Swanson, 1938, says something to the effect that it was well known in Naval Intelligence that Earhart had been picked up by the Japanese (check exact quote).

February 19, 1938 per New York Times, (conflict with January 13 date above) Vincent Astor sails from NYC aboard Panama Pacific liner *Pennsylvania* for Panama to board his yacht *Nourmahal* (in reality for espionage cruise to Marshall Islands at FDR's request). Astor and Kermit Roosevelt aboard (Roosevelt may have boarded in Honolulu?). Ship is not allowed by the Japanese to enter port in Marshalls and visits Gilbert and Ellice Islands, Samoa, and Fiji. It gathers information which Astor reports back to FDR by handwritten letter in May 1938 hand carried to FDR by Kermit and to the Office of Naval Intelligence [19 page typewritten report not yet found]. Ship returns to Panama May 21, 1938, transits the next day to the Caribbean and then to Bermuda and back in NYC by May 29, 1938. Some wonder if George Putnam had some connection with the cruise of the *Nourmahal*, possibly getting aboard in the Galapagos Islands or Panama.

March 17, 1938 letter from Mr. Carle Heine was published in the *Pacific Islands Monthly Magazine* dated May 25, 1938 describing an unclaimed letter to Miss Amelia Earhart (Putnam), Marshall Islands (Japanese), Ratak Group, Maloelap Island (10), South Pacific Ocean that he found in the Jaluit Post Office in the Marshall Islands with a return address to the Hollywood Roosevelt Hotel, Hollywood, California. The letter was date stamped Los Angeles, California October 7, 10 p.m., [1937]. Margo DeCarie, Earhart's personal assistant had been living in the Hollywood Roosevelt Hotel during September-October 1937.

April 4, 1938 Mrs. Eleanor Roosevelt presents Harmon Trophy to Jackie Cochran as the outstanding female pilot of 1937.

Early April, Archbishop Spellman returns to Boston from South American cruise.

Late spring 1938, George Putnam and friends return to Los Angeles? from expedition to the Galapagos Islands in the yawl *Athene*.

May 21, 1938 *Nourmahal* with Vincent Astor returns to Panama from Marshall Island espionage cruise, transits Panama Canal the next day, goes to Bermuda and then returns to NYC May 29, 1938. Count Carlo Frasso, and Dr. Gordon Stevenson aboard, completing 21,000 mile cruise.

June 25, 1938 Fred Noonan is legally declared dead by request of his widow, Mary Bea Martinelli.

June till mid July (summer) of 1938 Mark Walker (later PAA first officer on the *Hawaii Clipper* when it was lost) told his nephew Robert Greenwood of sabotage on the new PAA Boeing 314 at Seattle (per Charles N. Hill).

July 10-14, 1938 Howard Hughes with 3 crew members sets around the world record in a Lockheed Super Electra (L14) departing and ending at Floyd Bennett Field, NY.

July 29, 1938 PAA *Hawaii Clipper* disappears without trace enroute Manila from Guam. Suggested hijack by Japanese per Charles N. Hill, *Fix on the Rising Sun*.

In 1938 Atlas Corporation (Floyd Odlum, Jackie Cochran's husband, was President), began buying shares of Northeast Airlines (originally started in 1933 by Amelia Earhart, Eugene Vidal, Paul Collins, and Sam Soloman) obtaining a majority share in the 1940s, and in 1944 Jackie Cochran became a member of the Board of Directors.

1939
January 9, 1939 Earhart is declared legally dead at Putnam's request.

April 24, 1939 (per Gannon) Francis J. Spellman was made the archbishop of NYC by Pope Pius XII. Spellman moves from Boston to NYC and makes his home there until his death in February 1967.

May 21, 1939 George Palmer Putnam marries Jean Marie Cosigny in Boulder City, NV. (his 3rd marriage).

August 19, 1939 is date of section of US State Department file referred to in September 7, 1946 US State Department file which says "Mrs. Putnam wishes the U.S. Government to henceforth consider her a national of the Nipponese Imperial Islands." [Gibson/Gervais].

November 1939 George Putnam wrote and published his biography of Amelia Earhart entitled *Soaring Wings*. Relatively little said about her disappearance.

December 1939 Lt. Col. Dwight Eisenhower leaves MacArthur's staff in the Philippines for assignments at military posts along the west coast of the U.S.

December 11, 1939 Pope Pius XII made Spellman the Vicar of the US Armed Forces overseeing spiritual services to all Roman Catholics in the the US military.

1940
Beginning in 1940 thru 1945, Floyd Odlum has series of high level jobs in government in Washington, D.C. including Special Advisor to Chairman of Office of Price Administration, Director of Office of Production Management.

1941
By 1941 Archbishop Spellman was listed as the Grand Protector and Spiritual Advisor of the Sovereign Order of Malta, which was heavily involved with the Catholic Church and intelligence worldwide. FDR proposed and Spellman accepted role as a clandestine agent for FDR in the far corners of the world. "He would carry messages for the President, present the American point of view forcefully when necessary and act as Roosevelt's eyes and ears." (Cooney). Spellman is an OSS agent during WWII as was Bernt Balchen in USAAC in Scandinavia.

April 20, 1941 Kelley named Rt. Reverend Monsignor Kelley by friend Pope Pius XII, whom Kelley had earlier taught English in Belgium.

July 11, 1941 Kelley received a citation and medal from Secretary of the Treasury Henry Morgenthau (head of US Coast Guard) for "3 years of patriotic service with integrity and diligence for the Treasury Department (includes US Coast Guard) of the USA (no other data found).

December 7, 1941, "Pearl Harbor." War between US and axis powers declared.

Possibly sometime in December 1941 the Secretary of the Navy Frank Knox, a friend, was in Kelley's office at Seton Hall College, South Orange, NJ.

December 15, 1941 Admiral Nimitz appointed Commander-in-Chief US Pacific Fleet. Nimitz arrives in Pearl Harbor Christmas Day. 1941 and pursues victory in Pacific.

Early in WWII Kelley was appointed Chaplain for the Atlantic Overseas Air Command at Newark Air Base, NJ through assistance of Archbishop Spellman and served in that role concurrent with being President of Seton Hall College (University) throughout the war. Kelley had a pilot's license for about 7 years [what airfield did he fly out of?] and said he actually flew a P51 out of Newark Air Base on one occasion in 1943 [perhaps he flew in the jump seat with a more experienced pilot?].

1942

June 15, 1942 Kelley was appointed Permanent Advisor to Mt. St. Dominic which had become a college in 1939, and which Irene O'Crowley (Craigmile Bolam) had attended in 1916 in the 7th grade.

August 29, 1942 Putnam enters Army Air Corps as Captain, completes Army Air Force Officer Training School in Miami Beach, FL October 17, 1942 and Intelligence School in Harrisburg, PA early in 1943.

September 1942 Cochran appointed head of USAAC Women Pilot's Training and shortly later as head of the WASP by General "Hap" Arnold (she and WASP were civil service, not military).

November 1942 Putnam writes and has published his autobiography, *Wide Margins* . Very little said about Earhart's disappearance.

February 9, 1943 Archbishop Vicar General Spellman departs NYC on 24 week trip to Europe, the Middle East, Africa, South America visiting US military bases and people and Catholic entities. Returns to NYC August 1, 1943.

Early in 1943 Captain George, Putnam USAAC, graduates from USAAC Intelligence School in Pennsylvania, is promoted to Major, and is assigned to B29 units at Smoky Hill Army Airbase, Salina, KS.

April 1943 RKO Studios owned by Odlum and Cochran produces and distributes film *Flight For Freedom* with Rosalind Russell and Fred MacMurray in 1943. George Putnam had developed first film script for the film in 1939, which was modified by others. Cochran reviewed the script and Odlum and Cochran followed the production of the film closely.

January 31, 1944 American invasion of Marshall Islands (part of Operation Flintlock). American military personnel (including Captain later Vice Admiral Edgar A. Cruise, per Goerner) start receiving information from natives that a Caucasian man and a woman (flyers) came down in the Marshalls and were picked up by the Japanese and taken to Saipan.

January 31, 1944 Commander Richard B. Black USNR involved in invasion of Kwajelein.

March 4, 1944 an AP report by war correspondent Eugene Burns published in the *New York Daily News* March 22, 1944 (and in other newspapers) reports a mission-trained native Eli [Jibimbam] told USN officers then stationed in the Marshall Islands, Lt.j.g. William J. Bauer (of Lone Pine, CA), Lt. Eugene T. Bogan (of NYC and an attorney), and Lt. James B. Toole (of Washington, DC) that he had been told by a Japanese trader named Ajima three and a half years ago that "an American woman pilot came down about 1937 between Jaluit and Ailinglapalap Atolls and that she was picked up by a Japanese fishing boat and the trader Ajima heard that she was taken to Japan."

April 1944 Putnam is Major, intelligence officer, in 20th Air Force, 58th Bomb Wing, 468th Bomb Group (assigned to the 792nd Bomb Squadron for logistical support) leaves Smoky Hill Army Airfield and flies in B29 via Newfoundland, Marrakesh, and Cairo, Egypt to Kharagpur, India in CBI, as part of "Operation Matterhorn." He goes to Chengtu, China in June 1944 with his units. He briefs group for first B29 raid against Japan for the June 14, 1944 raid against the Yawata Steel Works on Kyushu. He also spends much of his free time looking into clues and stories about Earhart disappearance (Morrissey 1963).

Sometime in 1944 Putnam makes a 3 day trek behind enemy lines to listen to a Japanese radio broadcast from a US Marine Corps (SACO) listening post near the China coast to hear a female Japanese radio voice, which Putnam denies is his former wife (Morrissey).

June 15, 1944 American invasion of Saipan, part of "Operation Forager." American military personnel start receiving information from Saipanese natives that a Caucasian man and woman (flyers) arrived on Saipan [either in their own plane or more likely in a Japanese seaplane] about 1937, were kept prisoner, and later were killed or died, or that Earhart survived and was taken to Japan.

Commander Richard B. Black involved in invasion of Saipan at Charan Kanoa as Beachmaster and part of landing party.

June 25-26(?), 1944 Sgt. Thomas Devine, 1st Sgt. 244th US Army Postal Unit, says he saw Earhart's Electra at Aslito Airfield fired upon by US military aircraft and burned under the direct control of Secretary of the Navy, James Forrestal who was present.

Lt. Col. Wallace Greene USMC is operations officer for 2nd Marine Division on Saipan, identified Earhart's aircraft on Saipan at Aslito Airfield, and knew about orders to burn Earhart aircraft according to Sgt. Erskin Nabers USMC radio operator on Saipan at the same time and who recalled to Sgt. Thomas Devine that he had received and read those messages which were delivered to Col. Greene. Green later in January 1964 to 1967 became Commandant of the Marine Corps. In 1965 while Commandant of the Marine Corps. Greene was interviewed by Goerner and others, but made denials about his knowledge of Earhart or her aircraft. Admiral Nimitz told Goerner that apparently the Marine Corps was covering up their knowledge of Earhart and her aircraft.

General Alexander A. Vandegrift, Commandant of the US Marine Corps, states in letter about 1963 to Fred Goerner that he was told by USMC General Thomas Watson in 1944 (then commander of 2nd Marine Division on Saipan) that Earhart and Noonan had perished on Saipan.

USMC General Graves B. Erskine, then Chief of Staff of V Amphibious Corps under USMC General "Howling Mad" Smith during the invasion of Saipan, in intelligence capacities, said "We did learn that Earhart was on Saipan and that she died there" according to Fred Goerner.

June 19, 1944 United States Military Government in the Northern Mariana Islands is established by Proclamation No. 1; Fleet Admiral Chester W. Nimitz becomes Military Governor of the Mariana Islands. He remains so until November 24, 1945 when he is succeeded by Admiral Raymond Spruance.

August 1944 Jinx Falkenburg (Hollywood star and model) and Lt. Col. Tex McCrary USAAC, visit Chengtu, China on USO tour. Likely Major George Putnam was there at the same time.

August 29, 1944 Major General Curtis LeMay arrives at Khragpur, India and takes command of the 20th Air Force Bomber Command.

Sometime after the invasion of Saipan in June 1944 and before December 1944, Major George Putnam visited Saipan investigating rumors of Earhart having been on Saipan. Supposedly he found no evidence.

Late fall, early winter of 1944 Major George Putnam returns to US across the Pacific in the sick bay of a troop transport ship.

In 1944 Jackie Cochran became a member of the Board of Directors of Northeast Airlines. In 1938 Atlas Corporation (Floyd Odlum, Jackie Cochran's husband, was President), began buying shares of Northeast Airlines, originally started in 1933 by Amelia Earhart, Eugene Vidal, Paul Collins, and Sam Soloman. Atlas Corporation obtained a majority share in the 1940s.

1945
January 15, 1945 last B29 raid out of China, all B29 units pull out of China within short time thereafter.

March 2, 1945 George Putnam divorced in N. Hollywood, CA. by Jean Marie Cosigny while George was recuperating from kidney infection acquired in China.

April 12, 1945 FDR died in Warm Springs, GA (age 63) succeeded by Harry Truman.

May 8, 1945 "VE Day," Victory in Europe, WWII

June 10, 1945 George Putnam (4th marriage) marries Margaret (Peg) Haviland in San Marino, CA.

June 10, 1945 "Tex" McCrary marries famous star and model Jinx Falkenberg in NYC.

Msgr. Kelley's brother is a Commander in US Navy, (duty and locations unknown as of 2/3/04).

Mid-July 1945 Sgt. Thomas Devine of 244th US Army Postal Unit says he was told by a Saipanese (Okinawan) female of Earhart and Noonan's execution and shown burial site near Garapan on the island.

August 9, 1945 Jackie Cochran departs San Francisco via military transport aircraft to Honolulu, Kwajelein, Marshall Islands, and Guam. She is on a mission for General Arnold as Special Consultant with priority #1 orders and is also a war correspondent for *Liberty Magazine.* A. Earhart had written an article for them in February 1937 before her disappearance and George Putnam wrote several articles for them in the couple of years after her disappearance. Cochran enroute visits Commodore Ben Wyatt, Commander of the Marshall Islands who was in navy intelligence? and who had known? Earhart. Wyatt probably visited with Vicar General Spellman as he passed thru a couple of days later.

August 12, 1945 Archbishop Spellman departs via air from NYC, destination the Pacific via San Francisco.

August 15, 1945, WWII ends (VJ Day). Spellman departs Honolulu for Kwajelein and Eniwetok, Marshall Islands. Continues on to Guam, where he spends about a week and is hosted by Jackie Cochran and Admiral Nimitz.

August 16, 1945 General Wainwright is liberated from POW camp in Manchuria and is flown to treatment and assembly camp in Korea.

August 17, 1945 An OSS team as "Operation Duck" with USA intelligence officer Lt. Jim Hannon parachuted into Weishein Civilian Internment Camp in Northeast China to liberate the camp as the first such liberation in the entire Far East. Hannon found a "Lady Yank" almost comatose in the Japanese portion of the camp. [Earhart would have been 48 years old]. Some researchers say the "Lady Yank" was Amelia Earhart and she was flown (approximately Aug 31 Sep 34 per Jim Hannon) from Weishein, China to a special camp in Korea where other important American POWs were first taken for assessment and preparation for return to the United States. Such VIPs included General Jonathan Wainwright, formerly General Douglas MacArthur's deputy in the Philippines who was captured by the Japanese and kept in POW camps in the Philippines, China, and in Manchuria until liberated August 16, 1945. Wainwright attended the Peace Signing aboard the Battleship Missouri in Tokyo Bay on September 2, 1945 and the surrender of Japanese General Yamashita ceremony at Baguio in the Philippines on September 3.

About August 18-26, 1945 Cochran, Spellman, Nimitz (then ranking Navy officer in Pacific), General Barney Giles (then ranking USAF officer in the Pacific as deputy to General Arnold), General "Tooey" Spaatz, General LeMay?, General Doolittle, and other flag officers are all on Guam. Cochran played poker with many of them.

August 26, 1945 Spellman flies to Manila from Guam at the request of General MacArthur and met McArthur there, the day before MacArthur's departure for Japan(August 27, 1945).

August 27, 1945 US Navy begins landings in Tokyo Bay area.

August (?) September 1945 to January 1946, Commander Richard B. Black involved in invasion and occupation of Japan.

August 28, 1945 first American troops airlanded at Atsugi Airfield, southwest of Tokyo, under command of Col. Charles P. Tench.

Lt. Col. "Tex" McCrary enters Japan before MacArthur (August 30, 1945).

August 30, 1945 beginning at 6 am a military transport airplane landed about every 34 minutes at Atsugi Airfield. General MacArthur lands about 2 p.m. at Atsugi Airfield and makes his temporary headquarters in Yokohama. On September 9, 1945 he makes his headquarters in Tokyo.

Rear Admiral Richard Byrd entered Japan on special assignment and was transported (on different dates?) by 318th Troop Carrier Squadron as was Archbishop Spellman.

August 31-September 34 (approximately, per Jim Hannon) "Lady Yank" flown from Weishein, China via Tsingtao to Korea. Others then suggest she was flown disguised as a nun via Japan, to USA.

August-September, 1945 some suggest McCrary met with Archbishop Spellman and Jackie Cochran in Japan and possibly China and had hand in arranging repatriation of "Lady Yank" (approximately Aug 31 Sep 34 per Jim Hannon) from Weishein, China to USA.

September 2, 1945 "Peace Signing" takes place on *USS Missouri* in Tokyo Bay.

September 3, 1945 Cochran met Wainwright at Baguio, Philippines at General Yamashita's surrender. She and war correspondent Shelley Mydans (former civilian internee in the Philippines) had flown to Manila earlier with Major General Barney Giles, deputy to General "Hap" Arnold. Cochran met Spellman again while she was in Manila. Cochran, Shelley Mydans, and her husband Carl Mydans, also a war correspondent/photographer civilian internee in the Philippines and China, were later together in Japan and flew in Cochran's C54 from Japan to China together.

September 8-9, 1945 (Gannon) Spellman landed at Inchon, Korea, drives in Jeep to Seoul with Admiral Barbey, and attended the Japanese surrender in Korea (Seoul) at invitation of Admiral Barbey (who was the senior officer of Captain Wagner who had loaned Jackie Cochran a PBY in the Philippines about the same time). Spellman had departed from the Philippines and had flown to Okinawa where he had remained for a couple of days being held up by a typhoon before going on to Korea.

September 12, 1945, two days later, (the same day as General/Prime Minister Tojo's attempted suicide) Spellman flew via Okinawa to Japan with 318th Troop Carrier Squadron, landing at Atsugi Airfield near Tokyo very early in the occupation. Admiral Richard Byrd also was transported by this same squadron (at a different time?).

September sometime between 12-18th, 1945 Archbishop Spellman was first important foreigner to be received by Emperor Hirohito in Tokyo with General MacArthur's permission. From then on MacArthur and Spellman were close friends. Spellman was well connected to all American Presidents until his death.

September 18(?), 1945 Jackie Cochran after flying back to Guam from the Philippines with General Giles, also lands in Japan very early as the first American woman to enter Japan after the end of the war. (Cochran met McCrary in Japan about Sep 19 per Joe Klaas). She entered the Dai Ichi Building (Japanese military aviation headquarters in Tokyo) where she found files on Earhart and herself among others, but those files have never been seen since by western investigators. She was traveling on assignment with Priority #1 orders from Army Air Forces Chief, General "Hap" Arnold and as a *Liberty Magazine* war correspondent. After she arrived in Japan she was provided her own C54 aircraft and crew for movement in the Far East by Army Air Corps General George Kenney. Spellman had departed Japan about the 18th? for China via Okinawa in a C46.

September 18, 1945 Generals Giles, LeMay, and "Rosie" O'Donnell fly in a B29 nonstop from Hokkaido, Japan to U.S. (Washington, D.C.?).

According to researcher Dean Magley, Amelia Earhart assumes name of Irene Craigmile while she is in Japan shortly after the end of the war and then is air transported to New Jersey with the help of the Catholic Church and the American military. Suggestions have been made that Kelley was asked at the end of the war by Archbishop/Cardinal Spellman to help a repatriated Amelia Earhart recover from her captive life under the Japanese, and assume a new identity in the USA. Dean Magley met with Kelley at least one time and corresponded with him in the 1980's and recorded rather convincing discussions with Kelley confirming this.

September 18, 1945 (approximately) Spellman departed Japan in a C46 for China via Okinawa. On or about the 18th19th? he stayed with Lt. General Wedemeyer in Liuichow, China and then went on to Kumming. He was in Calcutta on September 26 and returned to NYC October 3, 1945 (per Gannon), later (per Jackie Cochran).

September 27, 1945 General MacArthur has first visit with Emperor Hirohito in Tokyo.

October 6-7 (?), 1945 Cochran flies, with Shelley and Carl Mydans, to China from Japan in personal C54 assigned by USAAC General Kenney, met with her husband Floyd Odlum's cousin, Canadian Ambassador to China, General Victor Odlum, senior American military officers, and other dignitaries.

October 10, 1945 General Jonathan Wainwright awarded Honorary Degree by Seton Hall and Msgr. Kelley, and stayed at Seton Hall for a time. Recall that Wainwright had been brought to an assembly camp in Korea for VIPs (when he was first released from a POW camp in Manchuria), where some have suggested Earhart was taken from Weishein. Also Cochran met Wainwright at Baguio, Philippines at the General Yamashita surrender on Sep 3, 1945.

Mid November, 1945 Archbishop Spellman (per Cochran in *The Stars at Noon*; conflict of date with the book *The Cardinal Spellman Story* by Gannon which says Spellman returned to NYC October 3, 1945). Cochran, and Mrs. Curtis (Helen) LeMay, wife of Major General Curtis LeMay, former commander of 20th Air Force Bomber Command in China and later commander of the 21st Air Force Bomber Command operating from the Marianas Islands, [which was heavily bombing Japan January-August 1945], meet Spellman in Rome according to Cochran (per Gannon, Spellman was already back in NYC). Cochran has audience at Vatican with Pope Pius XII (friend of Archbishop Spellman and Mgsr. Kelley).

McCrary returns to NYC as civilian after the war and is friends with and visits Irene Bolam occasionally. He is also friends with Jackie

Cochran who lives part time in NYC. Cochran and Odlum are friends with Eleanor Roosevelt as well, and were big backers of Truman about this time (19451951) according to Klaas (*Amelia Earhart Lives*), until backing Eisenhower in 1952.. Cochran and McCrary played major role in his acceptance as Presidential candidate in 1952. Eisenhower had been a deputy to MacArthur in Washington and the Philippines in the 1930s.

Kelley was friends with President Truman, his wife, and daughter Margaret. He was also friends with Charles Lindbergh and family, Fred MacMurray and Rosalind Russell (stars of 1943 film *Flight For Freedom*), Katherine Hepburn, and Archbishop/Cardinal Spellman among many other notables.

Supposedly "Operation Stitch", an investigation into Earhart's status, took place in Japan about this time under the control of the 441st CIC army intelligence unit in Tokyo. Those records have not been found by researchers.

November 20, 1945 General Dwight Eisenhower appointed Chief of Staff, US Army in Washington, DC by President Truman, through June 7, 1948, when Eisenhower became President of Columbia University, NYC.

November 20, 1945 Admiral Nimitz appointed Chief of Naval Operations by President Truman. Nimitz is retired from command, December 1947, and retires to Berkeley, CA. Remains active with US Government and UN assignments thereafter.

November 24, 1945 Admiral Raymond Spruance is named Military Governor of the Marshall, Caroline, and Mariana Islands, replacing Fleet Admiral Chester W. Nimitz.

1946
February 3, 1946 Admiral John H. Towers is named Military Governor of the Marshall, Caroline, and Marianas Islands, replacing Admiral Raymond Spruance, and remains so until succeeded by Admiral Louis Denfield, February 28, 1947.

February 18, 1946 Archbishop Spellman made a Cardinal by Pope Pius XII.

May 2-August 31, 1946 United States Commercial Company conducts economic survey of Micronesia. Commander Richard Black is leader).

July 1, 1946, 1st United States atomic bomb test in Marshall Islands, at Bikini

July 4, 1946 Inhabitants of Northern Mariana Islands permitted to reside outside centralized camps for first time since U.S. invasion of islands; United States grants independence to Philippines.

September 7, 1946 a US State Department file reference is made to a part of the file which says "Mrs. Putnam wishes the U.S. Government to henceforth consider her a national of the Nipponese Imperial Islands and that section is dated August 19, 1939." [Gibson/Gervais].

September 1946 Elizabeth Vining, went to Japan where she became an English tutor to the children of Emperor Hirohito from October 1946 till October 1950. She visited MacArthur 7 times at his Dai Ichi office during her time in Japan. In Fall of 1917, Earhart met Elizabeth Vining, a Philadelphia Quaker who was a teacher? (age 15?) at O'Gontz School (a finishing? school) in Rydal, N. Philadelphia suburb. Ms. Vining resigned about that time because of dissatisfaction with head mistress and owner of school. Ms Vining was born October 6, 1902 in Philadelphia and died November 28, 1999 in Philadelphia (age 97). She was a pacifist as Earhart was.

1947
February 28, 1947 Admiral Louis E. Denfield is named Military Governor of the Marshall, Caroline, and Mariana Islands, replacing Admiral John H. Towers.

May 1947 Fleet Admiral Chester Nimitz received an honorary degree from Seton Hall from Msgr. Kelley. In 1965 Nimitz was quoted by Fred Goerner as having said "Earhart had been picked up by the Japanese and taken to the Marshall Islands" and that Washington (U.S. Government) had the records of her survival.

July 18, 1947 President Truman approves Trusteeship Agreement, acting pursuant to joint resolution of Congress of same date; Truman issues Executive Order 9875, placing Trust Territory of the Pacific Islands under U.S. Department of Navy; Admiral Louis E. Denfield, Military Governor of the Marshall, Caroline, and Mariana Islands, is named first High Commissioner of the Trust Territory; Government of the Trust Territory of the Pacific Islands issues Proclamation No. 1 for the government of the Trust Territory

1947

April 23, 1948 Cardinal Spellman departs NYC for Australia, SE Asia, and China and returns to NYC in 52 days, about June 15, 1948.

June 7, 1948 General (Ret) Dwight Eisenhower becomes President of Columbia University, NYC thru December 16, 1950; resigns from Columbia and goes to Europe as head of NATO till candidacy as President of US in June 4, 1952.

Summer/Fall 1948 Floyd Odlum campaigns and raises money for President Truman presidential campaign.

1948

March 2, 1949 Kelley resigned from Seton Hall (then 13,109 students, 11 departments in undergrad school and 2 grad divisions, one in Newark, and one in Jersey City) and spent a little over a year at St. Patrick's Cathedral in Newark, NJ relaxing, writing, with no major church assignments.

July 24, 1949 in article in *Los Angeles Times* Mrs. Amy Earhart is quoted as saying "I am sure there was a government mission involved in the flight, for Amelia explained there were some things she could not tell me." "I am equally sure she did not make a forced landing in the sea. She landed on a tiny atoll one of many in that general area of the Pacific and was picked up by a Japanese fishing boat that took her to the Marshall Islands, then under Japanese control." "I know she was permitted to broadcast to Washington from the Marshalls, because the officials on the island where she was taken I can't recall the name of it believed she was merely a transocean flier in distress." But Tokyo had a different opinion of her significance in the area.

She was ordered taken to Japan. There, I know, she met with an accident, an 'arranged' accident that ended her life." Her story of Amelia's death, she said, is based on a mosaic of reports, letters, conversations bits of evidence and documentation that have built into an explanation which the mother believes implicitly.

August 1949, CIA G2 document on Earhart. U.S. government had pressed Japanese for as much information as they could obtain. American intelligence agents were unable to find any Japanese Navy records pertaining to Earhart. But according to Japanese navy personnel who had supposedly searched for the Electra, the Japanese Navy's 12th Squadron, assigned to the Marshalls in 1937, was instructed by Tokyo, after a request from the U.S. government, to send the *KAMOI*, a seaplane tender, and several large flying boats, using the sea to the south of Jaluit as a central search point. Later the survey ship *KOSHU* was ordered into the area. The Japanese testified that the *KAMOI* led the rescue effort, but no traces of Earhart were found. The investigation was closed. However research by Loomis and others found that the *KAMOI* was docked in Saipan on July 2 and left July 4 for Ise Bay, Japan where it docked on July 10. Clearly the Japanese government had lied to the U.S. in 1937 and again in 1949.

1950
January 4, 1950 George Putnam dies in Trona, CA (age 63) of uremic poisoning from parasitic kidney infection acquired in China in 1944.

May 31, 1950 Kelley was assigned as Pastor of the Mt. Carmel Church in Ridgewood, NJ and retired from that position May 31, 1976. He had acquired "Chez Nous", a 30 room, 12 bedroom mansion at Rumson, NJ in 1950. Kelley's family had considerable money and he had visited and had owned homes in several islands in the Caribbean over the years. In his memoirs he says he was "finally given a very nice home" on St. Croix, U.S. Virgin Islands by Mrs. Fairleigh Dickenson, which he called "My Early Heaven." Researcher Matt Rodina says Kelley purchased a home on St. Croix in 1969.

Commander Richard Black (with comparable rank of Colonel) enters Korea with 8th Army, retreats with the Army, and goes to Japan for "research."

December 16, 1950 Eisenhower becomes NATO Commander till retiring May 31, 1952 from US Army to become Presidential candidate.

1951
April 1951 General MacArthur recalled from Korea by President Truman and retired from Army. Began living in Waldorf Astoria Hotel in NYC with his wife until his death in 1964.

June 29, 1951 President Truman issues Executive Order 10265, transferring administration of Trust Territory of the Pacific Islands from U.S. Department of the Navy to U.S. Department of Interior, and Executive Order 10264, making the same transfer of administration for American Samoa.

1952
February 8, 1952 rally for "Ike Eisenhower for President" in Madison Square Garden arranged by Cochran, Tex McCrary and Jinx Falkenberg among others. February 10, 1952 Cochran sees Ike in Paris and shows video of the rally after flying to Paris immediately after the rally.

June 4, 1952 Eisenhower announced his candidacy for the Presidency after very big role played by Cochran and McCrary in securing his candidacy. Cochran and Odlum become good personal friends and prime backers of Eisenhower (possibly earlier than 1952).

November 4, 1952 Eisenhower nominated for Presidency. MacArthur was also a candidate in this election, but lost to Eisenhower and others. Eisenhower had been MacArthur's chief aide in Washington, D.C. beginning in 1933 and in the Philippines in 1935 to 1939.

November 10, 1952 President Truman issues Executive Order 10408, to return administration of Saipan and Tinian from U.S. Department of Interior to U.S. Department of the Navy.

1953
January 20, 1953 Eisenhower became President succeeding Truman and served to January 20, 1961. Died March 28, 1969. He spent much time at the Indio ranch of Floyd Odlum and Jackie Cochran, as did General and Mrs. Curtis LeMay.

Captain Richard Black USNR begins formal association with Office of Naval Research.

1957 March 12, 1957 Admiral Richard Byrd dies in Boston.

1958
Guy Bolam married Irene Craigmile [formerly Irene Craigmile Heller] in 1958 in Rye, NY (where George Palmer Putnam was born and lived and where he and Amelia Earhart lived into the mid 1930s). Another Irene

Craigmile was present in the greater NYC area per Ann Pellegreno. Bolam's marriage license was obtained in Greenwich, CT. Putnam/Earhart license was obtained in Noank, CT in 1931. Guy Bolam worked as an account executive for J. Walter Thompson in London from 1937-1940. Some say he was a British M16 agent, perhaps after 1940. The Guy Bolam's settled in Bedford Hills, NY and in 1967 they moved to Jamesburg, NJ.

October 9, 1958 Pope Pius XII died in Rome, succeeded by Pope John XXIII.

1960
March 1960 Lt. Col. USAF Paul Briand publishes *Daughter of the Sky* . Briand has Earhart and Noonan flying to Saipan (erroneously?), in her airplane, where they die or are killed. Major Joe Gervais USAF and Lt. Col. Robert Dinger USAF team up with Briand to form "Operation Earhart" to research Earhart's disappearance.

John Bolam (Guy Bolam's 30 year younger half brother) and wife Irene Bolam heard of Irene Craigmile Bolam being decorated by NASA in the 1960's and of her knowing some astronauts. USMC astronaut Wally Schirra in October 1979 tells researcher Magley that AE was alive in last couple of days.

1961
January 20, 1961 Eisenhower completed 2 terms as President, succeeded by Kennedy.

August 1961 Howard Hughes buys into Northeast Airlines from Atlas Corp. (F. Odlum owner). Odlum and Jackie Cochran had bought into Northeast Airlines earlier and Cochran was on their Board of Directors beginning in 1944.

1962
Commander John Pillsbury USN, Admiral Nimitz's aide, told Fred Goerner "The Admiral (Nimitz) wanted me to say to you that he thinks you should continue the investigation, and I want to add to that, don't you give up. You're on to something that will stagger your imagination."

1963 *Courage is the Price* by Muriel Earhart Morrissey is published.

1964

General Douglas MacArthur dies in NYC. His widow was said to have occasionally received visits from Irene Bolam (the Bolam that Gervais and Klaas knew), even after her supposed death in 1982 (per Jerry Steigman, per Tod Swindell). McCrary was also visiting Irene Bolam at least up until 1982.

General Lucius D. Clay, former staff member to General Douglas MacArthur and Major Eisenhower in the Philippines in 1937, and former pre-NATO Commander is quoted by Fred Goerner as saying "There's more to the Earhart business than anyone suspected. I'm not a part of it myself, but I would like to see it told."

1965

August 7-8, 1965 Joe Gervais invited to Sea Spray Inn, Westhampton, Long Island, NY by Viola Gentry. Meets Mr. and Mrs. Guy Bolam and ID's Mrs. Bolam as Amelia Earhart. Guy Bolam tells Gervais he worked for Amalgamated Wireless. But according to ad in *Broadcast Magazine Yearbooks,* 19671970, he is head of Guy Bolam Associates with affiliate offices all over the world and lists as vice-presidents; "I.R.O'Crowley" [Mrs. Guy Bolam's surname with her aunt's initials, and "I.M.Craigmile," her nuptial name two marriages back before she became Mrs. Guy Bolam]. Guy Bolam is also said to have worked for Radio Luxembourg, which is known to have been involved in covert broadcasts of radio messages.

Dinger drops out of "Operation Earhart" and Joe Klaas, Lt. Col. USAFR joins the group of Briand and Gervais. Dinger on active duty and Klaas in the AF Reserve were stationed at Hamilton Air Force Base in Marin, County, CA. Gervais was then retired from the USAF and working as a supervisor for the Las Vegas, NV school district.

December 1965, Fleet Admiral Chester Nimitz told author Fred Goerner "Now that you're going to Washington, Fred, I want to tell you Earhart and her navigator did go down in the Marshall Islands and were picked up by he Japanese." "This knowledge was documented in Washington. He also said that several departments of government have strong reasons for not wanting the information to be made public."

1966

February 20, 1966 Admiral Nimitz dies on Yerba Buena Island, San Francisco Bay. February 24, 1966 he is buried in Golden Gate National, Cemetery, San Bruno, CA. Cardinal Francis Spellman attends and recites prayer.

July 4, 1966 President Johnson signs Freedom of Information Act passed by Congress. It did not apply to investigatory files compiled for law enforcement purposes. Over the next months and years it was implemented by the various government agencies. Several amendments made over the years.

Late 1966 Goerner publishes *The Search for Amelia Earhart*. Goerner has Earhart and Noonan picked up by the Japanese in the Marshall Islands and being transferred to Saipan where they die or are killed by the Japanese. Book is a best seller and generates much attention. Book includes some data and observations by Thomas Devine, previously a Sgt. in the 244th US Army Postal Unit on Saipan in WWII and later author of *Eyewitness: The Amelia Earhart Incident* published in 1987.

1966
Late June, early July, Ann Pellegreno retraces (approximately) Earhart's flight (30th anniversary). Publishes book *World Flight* in 1971 about the flight and discussion of theories of Earhart's disappearance.

July 4, 1967 US Navy file on Earhart classified for 30 years released.

Early July, 99s and other early flyers present ceremonies honoring Earhart at Smithsonian Institution in Washington, DC.

Mr. and Mrs. Guy Bolam move to Jamesburg, NJ from Bedford Hills, NY.

December 2, 1967 Cardinal Francis Spellman dies.

March 28, 1969 President Eisenhower dies at Walter Reed Army Hospital, Washington, DC.

Msgr. Kelley purchases home on St. Croix per Matt Rodina.

1970
May 6, 1970 Guy Bolam died in New York City Hospital.

November 7, 1970 *Amelia Earhart Lives* by Joe Klaas (Lt. Col. USAF Ret) featuring Joe Gervais' research is published and is highly publicized in a major press conference in Los Angeles. The book is pulled off the bookshelves in late December 1970 by publisher McGraw-Hill without telling Klaas and Gervais. Klaas and Gervais say in book that Howard Hughes may have sold plans of his H1 *Racer* to the Japanese and that it became part of the design of the Japanese *Zero* fighter. They receive threats and warnings from Hughes staff members.

November 10, 1970 Irene Craigmile Bolam strongly objects to Klaas book in press conference in NYC.

1971 Ann Pellegreno publishes book, *World Flight*.

Irene Rutherford O'Crowley (retired lawyer, born in 1896) lives in rest home in Highlands, NJ in 1971. She is aunt of Irene O'Crowley [Craigmile, Heller, Craigmile, Bolam] supposedly born 1904. Irene Rutherford O'Crowley handled legal affairs for Earhart in the 1930's. Claims she first met Earhart in Europe when she and niece [Irene O'Crowley, later Craigmile, Heller, Craigmile, Bolam] were traveling in Europe and Earhart had just flown solo across the Atlantic to Europe. Irene Rutherford O'Crowley claims her niece did not meet Earhart then or later.

Irene Rutherford O'Crowley's Pilot Publishing Company associates are Miss "Craig" and Miss "Mile," which together spell "Craigmile," the first of the younger Irene Bolam's three married names.

Amendments to the Freedom of Information Act had become effective providing for broad access to FBI records which previously had been severely limited.

May 26, 1971 Irene Bolam issues Summons to McGraw-Hill, Joe Klaas, and Joe Gervais re book *Amelia Earhart Lives*.

July 1971 Irene Bolam files law suit against McGraw-Hill, Klaas, and Gervais. Wins 6 figure settlement (per John and Irene Bolam?, about $60,000 per Swindell) from publisher McGraw-Hill, but nothing from Klaas and Gervais after she refuses to produce finger prints and refuses to describe where and under what circumstances she met Earhart and obtained her flying experience. McGraw-Hill lawyer still believed there was truth in the book, but it was too costly for McGraw-Hill to further fight the case.

1973
October 17, 1973 Bernt Balchen, dies in Mt. Kisko, NY. By this time he and Admiral Byrd were enemies by Byrd's choice because of Balchen's doubt of truthfulness of Byrd having flown over North Pole.

1976
May 31, 1976 Msgr. Kelley retires from Ridgewood, NJ Parish and moves to Rumson, NJ.

June 17, 1976 Floyd Odlum dies in Indio, CA (age 84).

Msgr. Kelley begins (maybe earlier in 1960s per Tod Swindell) telling friends, (Donald de Koster of Detroit and St. Croix and Mrs. Helen Barber of Valley Forge, PA and Saint Croix), of his involvement with Earhart/Bolam per 1991 telephone interview by Reineck with de Koster and Barber. Kelley's mental faculties begin to wane about this time.

1979
October 1979, astronaut Wally Schirra tells researcher Dean Magley that AE was "alive in last couple of days."

1980
August 7, 1980 Jackie Cochran dies in Indio, CA (age 74).

1982
June 18, 1982 Symposium on Earhart sponsored by the Smithsonian National Air & Space Museum held in Washington, DC.

July 7, 1982 Irene Bolam died [age 78] of cancer at Roosevelt Hospital, Edison, NJ, her residence was Rossmoor, NJ with the Jamesburg address. She had stipulated in a written agreement with Rutgers Medical School that there would be no fingerprints or DNA taken; no other identity made available to the public or family. She had also stated she did not know who her parents were. She donated her body to the University of Medicine and Dentistry of New Jersey in Newark. Eventually she was cremated and buried in an unmarked grave. Her death certificate originally filed signed by Dr. Ming Fong Hsu of Roosevelt Hospital in Edison, NJ quotes Mrs. Bolam's "surviving spouse, Guy Bolam," as the person supplying the personal information for the document. But Guy Bolam had died in 1970! The death certificate was corrected on August 13 at the request of Irene Bolam's attorney Judge Edward Kennedy of New York (also a professional associate of Irene Bolam's aunt, Irene Rutherford O'Crowley, an attorney). Kennedy said "It was a struggle to get them to do it," adding that "it was the undertaker's mistake" in the first place. The undertaker, Stanley J. Bonczek of Jersey City, NJ said "he got the information for the death certificate from the medical school." An attorney for the medical school said "that the University staff did not participate in the preparation of Mrs. Bolam's death certificate."

October 18-29, 1982 *Woodbridge News Tribune* , NJ Newspaper publishes 10 day series of articles about lives of Amelia Earhart and Irene Bolam and suggestions of being one and the same.

October 29, 1982 *Woodbridge News Tribune* Publisher John Burk is MC for memorial dinner (arranged by Irene Bolam before her death) held at Forsgate Country Club, NJ.

Letters from Amelia by Jean Backus published.

1985
Amelia Earhart, The Final Story by Loomis (Major, USAF Ret) and Ethell published.

1987
Amelia My Courageous Sister by Muriel Morrissey and Carol Osborne published.
Eyewitness: The Amelia Earhart Incident by Devine published.
The Earhart Disappearance, The British Connection by Donohue published.

1988
The International Group for Historic Aircraft Recovery TIGHAR initiated its study of Earhart's disappearance. From the beginning they have focused on Nikumaroro Island (formerly Gardner Island in 1937) in the Phoenix Islands group. To date (June 2004) they have uncovered no confirmed evidence, after 8 expeditions to the island.

1989
Amelia Earhart Research Consortium (AERC) formed.

July 2, 1989 AERC Symposium supported by and held at Western Aerospace Museum, Oakland,

September 30, 1989 AERC Symposium supported by and held at Western Aerospace Museum, Oakland.

November 35, 1989 AERC Symposium held at Purdue University.

The Sound of Wings by Lovell is published.
A Biography of Amelia Earhart by Rich is published.

1990
Amelia Earhart Society (AES) formed with Bill Prymak as President, Editor, and Publisher of their quarterly newsletter which continued until 2000.

1991 Rear Admiral Richard B. Black died.

1993 April 30, 1993 Earhart Symposium held at the US Naval Institute, Annapolis, MD

August 27-29, 1993 AES Symposium held at Flying Lady Restaurant at Morgan Hill, CA

1994
September 13, 1994 Fred Goerner died of cancer in San Francisco, CA (age 69).

Lost Star by Brink published.
Amelia Earhart, Lost Legend by Wilson published.

1995
Tuesday before September 14, 1996 Msgr. James Francis Kelley dies at Medical Center of Ocean County in Brick, NJ. (age 94).

Fall 1996 banquet honoring Linda Finch held at the National Air & Space Museum.

1997
January 18, 1997 Earhart Symposium at Western Aerospace Museum, Oakland.

June-early July Linda Finch retraces (approximately) Earhart's last flight (60th anniversary). Flight covered intensely by media, internet, and publications. Book published by Pratt & Whitney called *Amelia Earhart World Flight 1937, World Flight 1997 Linda Finch.* Banquets at Oakland arranged by Western Aerospace Museum held pre-departure and on return honoring Finch.

July 24, 1997, 100th Anniversary of Earhart's birth, Celebration at the Amelia Earhart Birthplace Museum, Atchison, KS then and every year since.

Whistled Like a Bird, The Untold Story of Dorothy Putnam, George Putnam and Amelia Earhart by Sally Putnam Chapman (George Palmer Putnam's granddaughter) is published.

Amelia: The Centennial Biography of an Aviation Pioneer by Goldstein and Dillon is published.

East to The Dawn: The Life of Amelia Earhart by Butler is published.

1998
March 2, 1998 Muriel Earhart Morrissey (A. Earhart's sister) dies (age 98) at Medford, MA. Author of *Courage is the Price* , 1963; and co-author with Carol Osborne of *Amelia, My Courageous Sister* in 1987.

1999
Amelia Earhart; The Mystery Solved by Long and Long is published.

January? Earhart Symposium held at Western Aerospace Museum
2000
AES continues as an email discussion group, earhart@yahoogroups.com, under leadership of Ron Reuther till Dec 2003, thereafter Michele Cervone.

2002
May 17-19, 2002 AES Symposium held at Western Aerospace Museum, Oakland, 65th anniversary of Earhart's last flight.

With Our Own Eyes; Eyewitness to the Final Days of Amelia Earhart by Campbell and Devine is published.

October 14, 2002 Margaret "Peg" Haviland Putnam Lewis (last wife of George Putnam dies in southern California. Supposedly had given all of her Putnam/Earhart papers to Sally Putnam Chapman (George Putnam's granddaughter). Supposedly those papers were given to Purdue University Library adding to their already major collection, now the largest collection of Earhart papers in the world.

2003

2003 (?) *Earhart's Flight into Yesterday* by Lawrence Safford (CAPT USN Ret), edited by Warren and Payne published.

July 29, 2003 "Tex" McCrary died in NYC in 2003 (age 93).

August 29, 2003 Jinx Falkenberg dies in NYC in 2003 (age 84).

December 2003 *Amelia Earhart Survived* by Rollin Reineck (COL USAF Ret) published.

Appendix II -

CAA Numbers on Amelia Earhart's Electra

Author Thomas Devine stated in his book "Eyewitness: The Amelia Earhart Incident," that he had written to the FAA and had received a response dated January 15, 1971[58], was informed of the following.

All aircraft in the U.S. are assigned a prefix letter of "N", a prefix which was allocated to the United States at the 1919 Paris Air Convention. If a plane does not cross a national boundary, the prefix letter need not be painted on the aircraft.

There were two secondary prefixes in use in the 1930s by the Civil Aeronautics Administration (then the name for Federal Aviation Administration). One was "X" for experimental craft, the other was "R" for restricted use, such as racing or long distance flights.

According to the 1971 FAA letter, the history of Earhart's Model 10 Electra was that it was assigned 16020 on a date not shown in their records. On 7-19-36, the craft was approved for an experimental license with the number NX16020. The following month on 8-7-36, the aircraft was approved for a restricted license with the number NR16020.

Following are my explanatory remarks regarding various numbers which have turned up in photographs of Lockheed Electras:

N16020	Identified in a number of early photos of the Model 10E bought by Purdue for AE; the plane she started with in her preparations for her round-the-world flight.
NR16020	Identified in later photos of the Model 10E Electra. Number changed after CAA approved "R" (Restricted) classification.
	Reassigned to a plane bought by Paul Mantz in 1946; Mantz owned it until mid 1961, when it was sold to a small aircraft investment company owned

---

[58] Devine, "Eyewitness: The Amelia Earhart Incident," pg 148.

	by three men; crashed on hillside in California near Bicycle Lake Army Airbase in late 1961.
R16020	This number turns up unexplainably in several photos:

- ♦ Identified in a photo classified for 28 years. It was located By Joe Gervais and determined to be the secret "XC-35". A copy was released to Gervais. The first flight of the XC-35 was on 5-7-37, approximately 3 weeks before AE took off on her round-the-world flight.

- ♦ Identified by Gervais in a photo of an Electra, which matched the paint job of ZK-AFD/1095/10A in another photo. (ZK-AFD owned by Union Airways, New Zealand.)

Appendix III -

Documents Relating to Amelia Earhart's Electra

LOCKHEED
Aircraft Corporation
BURBANK, CALIFORNIA
July 27, 1936

Bureau of Air Commerce,
Los Angeles Municipal Airport,
Inglewood, California.

Gentlemen:

We have sold Lockheed Electra Serial No. 1055 to Amelia Earhart, of 50 West 45th Street, New York City, and are enclosing herewith Bill of Sale duly executed by this corporation.

We are enclosing also application for re-assignment of license number NR-16020 as executed by Miss Earhart.

As soon as transfer of title to this airplane has been completed, please forward your Form AB-16, together with license card to Miss Earhart at the address given.

Very truly yours,

LOCKHEED AIRCRAFT CORPORATION

by *Cyril Chappellet*
Cyril Chappellet
Secretary

S
encl 2
CC to Miss Earhart

Letter from Lockheed to Bureau of Air Commerce re the sale of Lockheed 10E to Earhart

BILL OF SALE

KNOW ALL MEN BY THESE PRESENTS: That

    LOCKHEED AIRCRAFT CORPORATION,
      Burbank, California,

hereinafter designated as the Seller, for and in consideration of the

sum of - - - - - - - - - - - TEN & no/100 - - - - - - - - - - Dollars,

lawful money of the United States of America, and other good and valuable considerations, to it in hand paid by

      AMELIA EARHART,
     50 West 45th Street,
      New York City,

hereinafter designated as the Buyer, the receipt whereof is hereby acknowledged, does by these presents grant, bargain, sell and convey to the said Buyer, her executors, administrators and assigns, delivered at Las Vegas, Nevada,

   One Lockheed Electra monoplane complete,
   being Manufacturer's Serial No. 1055,
   Department of Commerce license number
   X-16020,

TO HAVE AND TO HOLD the same to the said Buyer, her executors, administrators and assigns forever. And the said Seller does for its successors and assigns covenant and agree to and with the said Buyer, her executors, administrators and assigns, to warrant and defend the title to the said property, goods and chattels hereby conveyed, against the just and lawful claims and demands of all persons whomsoever.

WITNESS our hand and seal this _22nd_ day of July, 1936.

        LOCKHEED AIRCRAFT CORPORATION
       by _____
          Agent

Bill of Sale for original Lockheed 10 Electra – Page 1

STATE OF NEVADA)
)
COUNTY OF CLARK)

On this 28th day of July, 1936, before me, A.H. Harrington, a Notary Public in and for the said County and State, personally appeared E. C. McLeod, known to me to be the agent of LOCKHEED AIRCRAFT CORPORATION, the corporation that executed the foregoing instrument, known to me to be the person who executed said instrument on behalf of the corporation therein named, and acknowledged to me that such corporation executed the same.

WITNESS my hand and official seal.   A.H. Harrington

My Commission Expires February 18, 1937.

Bill of Sale for original Lockheed 10 Electra – Page 2

Telegram from Bureau of Air Commerce to Earhart re authority to change the CAA number on her plane

GEORGE PALMER PUTNAM

2 West 45th Street,
New York City.
September 24, 1936.

Dear Carroll:

A.E. has asked me to acknowledge your telegram of September 21st received by her in Lafayette authorizing the NR license for her Lockheed Electra. This telegram will be posted in the ship and the "N" will be painted on out there.

I am venturing this letter to inquire if some further authorization or form is necessary ultimately to replace the telegram for display in the ship. If so, will you please send same here and we will do the necessary.

Sincerely,

GPP.

Carroll Cone, Esq.,
Aeronautics Branch,
Department of Commerce,
Washington, D.C.

Letter from GP Putnam to Bureau of Air Commerce responding to their telegram

Appendix IV -

Text of Gendarmerie Report of January 4, 1939

Message found in bottle
Soulac-Sur-Mer, France, October 30, 1938

Enclosure #9 to Despatch No 3590
Dated January 4, 1939
From the Embassy in Paris

NATIONAL GENDARMERIE

"Have been prisoner at Jaluit (Marshall) of Japanese in a prison at Jaluit. Have seen Amelia Earhart (aviatrix) and in another prison her mechanic (man), as well as other prisoners; held for so-called espionage of gigantic fortifications which are built at Atoll.

This thirteenth day of October nineteen hundred and thirty-eight at 9:15 p.m.

We, the undersigned, Felix DOURTH, the resident gendarme at Soulac-sur-Mer, department of the Gironde, dressed in our uniform, and in conformity with the orders of our superiors,

Appeared at our barracks:

BARRAT, Genevieve, wife of LAMBERT, 37 years old, housewife, residing at Soulac-sur-Mer, (Gironde), who declared to us:

"To-day, the 30th instant, at about 5 p.m., I was walking along the beach opposite 'La Pergola'. I saw a bottle floating on the waves.

Seeing that it was hermetically corked, I wanted to find out its contents. Inside there were three sheets of paper and a lock of hair.

Since one of the documents specified that the police should be notified, I am giving them all to you."

Read, approved and signed.

This bottle is of a content of 10 centiliters.

At the bottom of the glass it is marked as follows: V.B.2. It was closed with a cork and covered with wax.

The neck of the bottle is light brown.

The hair found in the bottle is light brown.

The inscription on the documents is as follows:

1. Further proof: a lock of hair.

1. "May God guide this bottle. I entrust to it my life and that of my companions in misery.

2. In ordinary handwriting:

3. RECTO:

"Have been prisoner at Jaluit (Marshall) of Japanese in a prison at Jaluit. Have seen Amelia Earhart (aviatrix) and in another prison her mechanic (man), as well as other prisoners held for so-called espionage of gigantic fortifications which are built at Atoll.

"Earhart and her companion were picked up by a Japanese seaplane and will be held as hostages, say the Japanese. I was a prisoner because I debarked at Mili Atoll. My yacht 'Viveo' sunk, crew massacred (3 Maoris), the boat (26 T) was supplied with wireless."

On reverse of paper:

"Having remained a long time at Jaluit as prisoner, I was enrolled by force as a bunker-hand on board 'Nippon Nom?' going to for Europe. Shall escape as soon as the ship is near the coast. Take this message immediately to the Gendarmerie in order that we may be saved.

"This message was probably thrown off Santander, and will surely arrive at the Vendee towards September or at the least October 1938, remainder in the bottle tied to this one, Message No. 6."

A third document, in shorthand, read:

"In order to have more chance of freeing Miss Amelia Earhart and her companion, as well as the other prisoners, it would be preferable that policemen should arrive incognito at Jaluit. I shall be with JO . . . eux and if I succeed in escaping . . . for if the Japanese are asked to free the prisoners, they will say that they have no prisoners at Jaluit. It will therefore be necessary to be craft in order to save the prisoners of Jaluit.

At the risk of my life, I shall send further messages.

This bottle serves as a float for a second bottle containing the story of my life and . . . empty, and a few objects having belonged to Amelia Earhart. These documents prove the truth of the story in ordinary writing and shorthand and that I have approached Amelia Earhart . . . believed to be dead.

"The second bottle doesn't matter.

"I am writing on my knees for I have only a little paper, for fingerprints taken by the police. Another with thumb."

"Message written on the cargo board, No. 6".

These objects were seized and handed to the Office of the Public Prosecutor.

The second bottle in question was not found.
1st copy to the Public Prosecutor at Lesparre.
2nd copy placed in the archives.
Done and sealed at Soulac-sur-Mer, October 30, 1938.

Appendix V -

Amelia Earhart Flight Covers

Amelia Earhart carried specially printed souvenir covers on all of her flights. Initially, they were carried in small quantities, strictly as souvenirs. During the 1928 Friendship Flight, for example, only three covers were carried, which were postmarked June 16, 1928 at Trepassey, Newfoundland and later on June 21, 1928 in London, England.

The number went up slightly with the 1932 solo flight across the Atlantic. On that flight, Earhart carried 50 covers, which bear a May 13, 1932 postmark in New York and a May 23, 1932 postmark in Londonderry, England. These were numbered and autographed.

On the 1935 Honolulu-Oakland flight, Earhart carried 49 numbered, autographed covers. They were cancelled January 11, 1935 in Honolulu and January 12, 1935 in Oakland.

On the 1935 flight to Mexico, Earhart carried 85 covers, 35 of which sported the 20 cent Mexican airmail stamp overprinted with "Amelia Earhart/vuelo/de bueno voluntad/Mexico/1935." This translated to "Good Will Flight 1935.[1] Unfortunately, only several hundred were printed, two thirds of which were secured by GP Putnam. The philatelic world was not pleased when they learned how few of the stamps would be available to collectors. The 1935 overprint is now a highly sought, expensive philatelic item. Occasionally reproductions come up for sale on the internet.

In 1936 the souvenir cover became a much more important financial factor, providing substantial funds for the round the world flight. Large quantities were sold and carried on the flight. They were sold through exclusive outlets negotiated by GP Putnam. The main outlet was the well-known and prestigious Gimbel's department store chain.

Ads would be placed in various newspapers and philatelic periodicals and orders forms would be conspicuously available in each store to make it easy for customers to purchase covers when they were shopping. The ads stated that the covers would be canceled at Oakland at the beginning and end of the flight, as well as enroute either in India or Australia. Also, a Howland Island cachet, the first in history, would be placed on the covers.

---

[1] "Amelia Earhart: A Postal Portrait", Ma rjory J, Sente, Scott's Monthly Stamp Journal, November 1977, Pg. 10-15, Vol. 58, No. 11.

The price of the 1937 round the world flight cover was $2.50, or $5.00 if signed by Amelia Earhart. A monograph written by the late Archbishop Spelling[2] reported that 10,000 covers were initially printed for the 1937 round the world flight. Of these, 6500 were sold and placed in the nose of Earhart's plane, before take off from Oakland to Honolulu on the first leg of the flight.

After the crash at Luke Field in March 1937, a second cachet was added to the covers reading, "Held over in Honolulu following Take-Off Accident of March 20, 1937." Another ad was placed by Gimbel's in philatelic periodicals in April 1937 indicating that a second cover was available for purchase for the second attempt at a world flight.

By the time she took off from Oakland in May, an additional 1000 covers were sold and delivered to Amelia Earhart to place on her plane. The second cover was essentially the same as the first one in appearance, except that in the lower left corner there was stamped "2nd TAKE OFF" in a small black box.

The only known surviving canceled cover from the first attempt at the round the world flight was one purchased by Mr. Elmer Dimity, who loaded the covers for the second takeoff on the Electra which Earhart received on May 28, 1937. After he had loaded the covers on the Electra, Mr. Dimity removed the one addressed to himself and placed it in his desk.

The only surviving covers from the second takeoff are those which Amelia Earhart gave to her mother before departing on the flight. A picture of one is shown in Muriel Earhart Morrissey's "My Courageous Sister."[3]

In addition to the covers sold by Earhart's representatives, there were covers issued by various organizations and cities to commemorate various events in Earhart's career. These included covers to commemorate Earhart's visit to various cities, as well as the 1928 transatlantic flight, the fifth anniversary of the 1928 transatlantic flight, the 1932 transatlantic flight, the annual Cleveland Air Races, and various other aviation events. One of the most famous of these events was a route inaugurated in 1929 between Cleveland Ohio and Detroit, Michigan.

In the years since Earhart disappeared, there has been a continued if not increasing interest in them by philatelic collectors. The most famous collector of Earhart philately was Archbishop Spellman of the Catholic Church, whose collection is now on exhibit in the stamp museum named

---

[2] "Amelia Earhart," Cardinal Spellman Philatelic Museum, Regis College, Weston, MA.

[3] Muriel Morrissey, "My Courageous Sister," pg. 47.

after him in Boston. Since Earhart's disappearance, there have been many other covers issued to commemorate her career and her last flight. Probably the first cover after her disappearance was created on the USCGC ITASCA. According to Dave Bellarts, the ITASCA cover is probably the rarest Earhart cover in existance next to the second attempt round the world cover.

Only a few were made for ITASCA crew members, and a lot of these have since been lost. The cover shown in the illustration is the one bought by Chief Leo Bellarts. It is definitely a one-of-a-kind cover, as Chief Bellarts had hand colored the inner circles of the design in light blue and ended up bringing the cover home to his wife, since Earhart had not of course picked them up at Howland Island.

There seems to be a thriving market for Amelia Earhart philatelic items these days, especially on internet auction sites.

Earhart souvenir cover order form – front

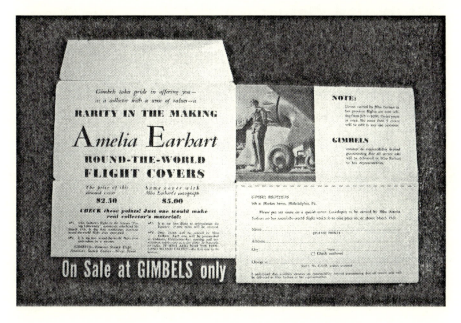

Earhart souvenir cover order form – inside

Round the World Cover (First Attempt) – Unused

Round the World Cover (First Attempt) - Cancelled

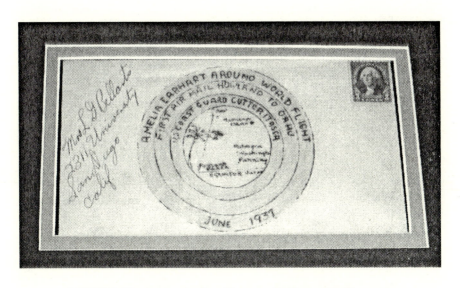

USCGC Commemorative Cover – Bellarts family copy

Appendix VI -

Summary of History of Paul Mantz' Lockheed Electra (Serial No. 1243)

The following is a summary of pertinent document copies I received from the FAA in August 2004. I have formatted this information in blocks, with a brief explanation of the activity that occurred, followed by a listing of the documents from which the information came.

It is for the reader to decide, but I could not find any evidence that this plane was connected in any way with Amelia Earhart.

Manufacture completed on Lockheed Electra Model 12A, Serial No. 1243, in November 1937. Lockheed applies to FAA for license for plane. Commercial license granted.

- Application by Lockheed for initial license [craft listed as being manufactured in November 1937] (11-29-37)
- Temporary License (12-4-37) – Number Assigned: NC 18955
- Permanent License (12-4-37) – Number Assigned: NC 18955

Plane sold by Lockheed on 12-5-37 to Western Air Express. Western Air Express applies to FAA for license. Commercial license granted.

- Bill of Sale for sale of craft by Lockheed Corp. to Western Air Express Corporation (WAEC) (12-5-37)
- Record, Transfer, and Reassignment Form (1-31-38) – title to WAEC
- Application by WAEC for license (12-4-37)
- License (12-15-38 Exp. Date) – Number Assigned: NC 18955

Plane sold back to Lockheed. Lockheed applies to FAA for license. Experimental license granted.

- Record, Transfer, and Reassignment Form (6-9-38) – Title to Lockheed
- Application by Lockheed for "experimental" license (4-25-39)
- Application by Lockheed for "experimental" license (5-1-39)
- License (11-15-39 Exp. Date) – Number Assigned: NX 18955
- Application by Lockheed for renewal of "experimental" license (12-2-39)
- License Authorization (12-23-39 Exp. Date)
- Application by Lockheed for Experimental License (6-20-40)
- Air Worthiness Authorization to Lockheed (8-29-40 Exp. Date)

Charles Babb Co., NYC buys aircraft. No record of license application by Charles Babb Co.

- Bill of Sale for sale of craft by Lockheed to Charles Babb Co. (6-24-40)

Charles Babb Co. sells plane to Canadian Department of National Defence. Plane redesignated N60775.

- Bill of Sale for sale of craft by Charles Babb to Canadian Department of National Defense for $50,250 (6-17-40)
- New number assigned on Bill of Sale: N60775

Canadian Department of National Defence sells plane to Algoma Air Transport Co. Ltd. Paul Mantz has already contracted with Edward Ahr Co. to purchase the Electra. Algoma immediately resells to Edward Ahr Co., who then immediately fulfills the earlier purchase agreement with Mantz. Eleven years later, in 1957, Paul Mantz requests the FAA assign Earhart's old number to his Electra for use of the plane in a film about Amelia Earhart.

- Bill of Sale by Edward Ahr Co. to Paul Mantz for $20,000 (1-21-46
- Bill of Sale by Canadian War Assets Corp. to Algoma Air Transport Co. Ltd. for $1 (1-24-46)
- Bill of Sale by Algoma Air Transport Co. Ltd. to Edward Ahr Co. for $1 (1-25-46)
- Bill of Sale by Edward Ahr Co. to Paul Mantz for $20,000 (1-25-46)
- Certificate of Ownership by Paul Mantz Air Services (5-15-46)

- Registration Certificate in the name of Paul Mantz Air Services (5-15-46)
- Letter from Paul Mantz Air Services to U.S. CAA requesting the assignment of Amelia Earhart's number (16020) to the aircraft

Paul Mantz sells the Electra to California Aircraft Investors in February 1961. In December 1961, the Electra is destroyed in a crash at Bicycle Lake. The FAA cancels the assignment of 16020 to this aircraft.

- Bill of Sale by Paul Mantz to California Aircraft Investors for $10 (2-14-61)
- Application for Registration by California Aircraft Investors (7-12-61)
- Certificate of Registration to California Aircraft Investors (8-9-61)
- FAA Aircraft Accident Notice [Indicating N16020 Canceled] (5-21-65)

The reader should note that there are many maintenance related documents not listed here, as well as a few license and air worthiness forms issued for various administrative reasons.

I determined through a written query to the FAA in the spring of 2004 that the registration number 16020 was reserved in 1987 at the request of Amelia Earhart's estate. This means that 16020 will never again be assigned to another aircraft.

Appendix VII -

ITASCA Radio Logs

What follows in the next pages are reproductions of the copies of the radio log pages provided to me by Dave Bellarts and copies of the corresponding pages from the national archives microfilm. They are presented here so that the reader can compare them and decide for him or herself about the discrepancies. Also, on the last page of this appendix is a copy of the Howland Island radio log of RM2 Cipriani's activities.

The reader should pay particular attention to the top of the first page of the genuine original preserved by Chief Bellarts in the right corner where the date is typed: "1 JULY 1937" in the blank reserved for the day and month. It appears in all caps. On the archive copy of that page, "1 July 1937" is typed in that blank, first letter only capitalized.

Moreover, the reader should notice that the genuine Bellarts copies are all typed in an all cap "communications" typestyle, whereas the first two pages of the archive copies are typed in an all cap standard typestyle. The third page of the archive copy is typed in al all cap "communications" typestyle, but there are differences in the way the date is typed in the upper right hand corner.

The archive copies are clearly not the documents donated by the Bellarts family.

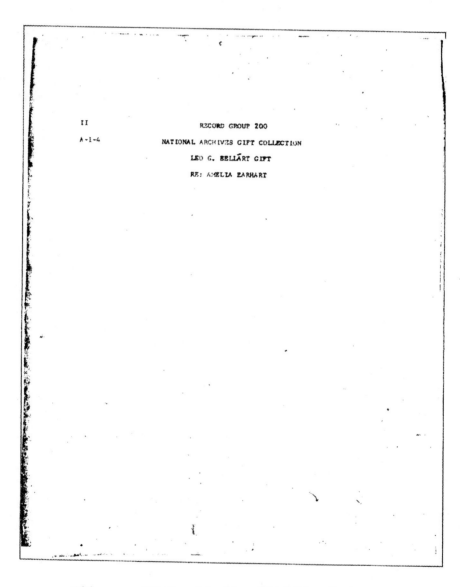

Title page of National Archives ITASCA radio log file

```
RADIO LOG  USCGC ITASCA                    Date   1 JULY 1937   193

L.G. BELLARTS CRM ON WATCH//    ENTRIES  3105 KCS                    TIME

WATCH STARTED AT// SIGNALS HEARD LOGGED ONLY//                       1900
VERY WEAK SIGS ON 3105 / UNREADABLE AND SEEMED SHIFT ABOUT           1917
NO SIGS ON DURING PERIOD                                             1945/48
WATCH REL'D BY GE THOMPSON RM3C         LG BELLARTS CRM              50
NO SIGS ON 3105                                                      2015-
KHAQQ DE NRUI A A A A A NRUI HOWLAND ISLAND A A A A A                2030-3
UNREADABLE FONE SIGS ON OR ABT 3105                                  35-40
SENT WEA TO KHAQQ ON 7500                                            55-58
KHAQQ DE NRUI A A A A A NRUI HOWLAND ISLAND A A A A A                2100-0
LISTENED FOR KHAQQ ON 3105 ND                                        15-18
SENT WEA 7500                                                        55-58
SENT WEA 3105/ KHAQQ DE NRUI AAAAA NRUI HOWLAND AAAAAAA              2130-3
UNREADABLE FONE SIGS ON OR ABT 3105                                  41-43
LISTENED FOR KHAQQ - SEVERAL CARRIERS BUT UNREADABLE                 45-48
GAVE WX TO KHAQQ 7500                                                55-58
KHAQQ DE NRUI A A AAAAAA NRUI HOWLAND ISLAND AAAAAAA                 2200-03
SEVERAL UNREADABLE CARRIERS ON OR ABT 3105                           03-07
PLANE UNHEARD                                                        15-18
KHAQQ DE NRUI SENT WEA 7500                                          28-29
KHAQQ DE NRUI AAAAAAA NRUI HOWLAND AAAAAAAAA                         30-33
LISTENED FER KHAQQ - SEVERAL UNREADABLE FONES BUT CANT MAKE OUT      45-48
GAVE WX TO KHAQQ ON 7500                                             25-28
KHAQQ DE NRUI AAAAAAAAA NRUI HOWLAND AAAAAAAAAA                      30-34
LISTENED FER KHAQQ BUT ND                                            45-48
SKNIXXEAXIXXMAXG SENT WEA TO KHAQQ                                   55-59
KHAQQ DE NRUI AAAAAA NRUI HOWLAND AAAAAAAA                           2300-0
LISTENED FER KHAQQ BUT ND                                            15-18
SENT WEA TO KHAQQ ON 7500                                            25-28
KHAQQ DE NRUI AAAAAAAAAA NRUI HOWLAND AAAAAAAA                       30-33
LISTENED FER KHAQQ ON 3105 ND QRM FONES ES UNREADABLE                45-48
KHAQQ DE NRUI (GAVE HER OUR WEA ON 7500)                             55-58
KHAQQ DE NRUI AAAAA NRUI HOWLAND AAAA PSE GA  3105 NOW K (UNHRD)     0000-0
KHAQQ UNHRD                                                          15-18
SENT WX TO KHAQQ ON 7500                                             25-28
KHAQQ DE NRUI AAAAAAA NRUI HOWLAND AAAAA PSE ANS 3105 (UNHRD)        30-34
KHAQQ UNHRD                                                          45-48
SENT WEA TO KHAQQ                                                    55-58
KHAQQ DE NRUI AAAAAAAA NRUI HOWLAND AAAAAAAAA                        0100-0
KHAQQ UNHRD                                                          15-18
SENT WEA TO KHAQQ                                                    25-28
KHAQQ DE NRUI AAAAAAA NRUI HOWLAND AAAAAAAAA                         30-34
KHAQQ UNHRD                                                          45-48
SENT WEA TO KHAQQ ON 7500                                            55-58
KHAQQ DE NRUI AAAAAAAA NRUI HOWLAND AAAAAAA                          0200-0
WATCH REL'D BY LG BELLARTS CRM      GE THOMPSON RM3C                 05
NOTHING HEARD ON 3105 KCS                                            0215-2
SENT WEA ON 7500                                                     28
AAAAAAAAA - AAAA NRUI HOWLAND                                        30-35
ITASCA TO EARHART / FONE 3105//                                      36
HEARD EARHART PLANE / BUT UNREADABLE THRU STATIC                     45/48
SENT WEATHER TO KHAQQ                                                0300
AAA AAA AAA ETC ETC TO KHAQQ/HOWLAND DE NRUI                         04
                        (OVER)
```

Original Bellarts ITASCA radio log 1-2 Jul 37 (First Page)

ENTRIES	TIME
PEPEATED WX ON FONE//	030
NOTHING HEARD FROM EARHART//	
SENT WX AT	15
A A A A A A DE NRUI / HOWLAND	30
REPEAT WEATHER ON FONE 3105//	34/
EARHART HEARD FONE / WILL LISSEN ON HOUR AND HALF ON 3105-SEZ SHE	35
BROADCAST WEATHER FONE 3105//	45/4--
REPEATED WEA ON KEY/ 3105 KCS/	0400
EARHART UNHEARD//	02
BRDSCT WX ON KEY AND FONE//	15/18
SENT WEATHER /CODE/FONE/ 3105 KCS--(HEARD EARHART -(PART CLSY)---	30/35
	53
EARHART UNHEARD//	0500
SENT WX AT- FONE KEY- 3105	15/18
AAAAAAAAAANRUI TO 7500 KCS// ((SETTING ON 3105 ALL TIMES//	30/35
NO HEAR DURING	40
SENT WEATHER / CODE AND KEY 3105 KCS/	45/50
WANTS BEARING ON 3105 KCS// ON HOUR// WILL WHISTLE IN MIC	0600-05
ABOUT TWO HUNDRED MILES OUT// APPX// WHISTLING// NW	14//
AAAAAAAANRUI (7500)-(LISTENING THRU ON 3105)	15
AAAAA NRUI --	30-
CALLED EARHART ON 3105 VOICE//	35
AAAAAAAAA NRUI NRUI TO //PSE ACKN 3105	36
PSE TAKE BEARING ON US AND REPORT IN HALF HOUR--	41
I WILL MAKE NOISE IN MIC -ABT 100 MILES OUT	45
AAAAAAAAAA 7500	46
AAAAAAAA - 3105	0705/
AAAAA NRUI ETC// 7500	08/12
AAAAAAA NRUI 7500 3105	12-14
FONE TO EARHART/ CANNOT TAKE BEARING ON 3105 VY GOOD/ PLESE SEND ON	14-16
500 OR DU WISH TAKE BEARING ON US / GA PSE / NO ANSWER	18
KHAQQ DE NRUI A-S GA 3105 / UNANSWD	19-24
KHAQQ FM ITASCA PLS GA ON 3105 KCS / UNANSWD / CRM/DC NOW ON THE	25
KHAQQ DE NRUI A-S GA 3105 / UNANSWD / NRUI'S MAIN D/F 500 KCS	26-9
KHAQQ FM ITASCA PLS Y OUR SIGS ON KEY PLS / UNANSWD	30
KHAQQ DE NRUI A-S	31-4
KHAQQ DE NRUI A-S, 7500 KCS	35-40
KHAQQ DE NRUI A-S, 3105 KCS	41
KHAQQ CLNG ITASCA WE MUST ON 00 YOU BUT CANNOT SEE U BUT GAS IS	
RUNNING LOW BEEN UNABLE TO REACH YOU BY RADIO WE ARE FLYING AT A	
1000 FEET	42
KHAQQ DE NRUI R MSG R QSA 5 R A-S (500 ES 3105) GA / UNANSWD	43-5
KHAQQ DE NRUI R MSG R QSA5 R A-S (3105)	47-9
KHAQQ FM ITASCA UR MSG OK PLS Y WID A3, A3/3105 / A-S FM NRUI A2	49-50-5
KHAQQ CLNG ITASCA WE ARE CIRHCLING, BUT CANNOT HR U GA ON 7500 VID A	
LNG COUNT LETTER N/ OR ON THE 3A3 TILL ON 4 HOUR (EPL.T 3 05)	53
KHAQQ DE NRUI A-S, 7500 / GA 3105	
KHAQQ CLNG ITASCA WE RECD WX UR SIGS BUT UNABLE TO GET A MINIMUM PSE	
TAKE BEARING ON US AND ANS 3105 WID VOICE / NRUI DE KHAQQ LNG DASHES	
ON 3105 / NRU12 DE NRUI P AR	0800-3
NRU12 GIVING TO DPE TT NO SIGS ON 3105 ES IMPOSSIBLE TO WRK ITA / R	04
KHAQQ FM ITASCA UR SIGS RECD OK WE ARE UNABLE TO HEAR U TO TAKE A	
BEARING IT IMPRACTICAL TO TAKE A BEARING ON 3105 OR VOICE HW DO U	
GET TT GA / UNANWD, 3105	05
KHAQQ DE NRUI GA ON 3105 OR 500 K, 7500	06

Original Bellarts ITASCA radio log 1-2 Jul 37 (Second Page)

```
Form 2611-A - Revised Oct. 1932
TREASURY DEPARTMENT
U. S. Coast Guard

RADIO LOG ............ ITASCA ............        Date ...... 2 JULY 1937

ENTRIES                                                                    TIME
W L GALTEN, SM3C ON — CGR-32-1
KHAQQ DE NRUI GA 500 WID DASHES ON 500 K / UNANSWD                         0805-8
KHAQQ FM ITASCA DID U GET TT XMISION ON 7.5 MEGS GA ON 500 KCS SO
  TT WE MAY BE ABLE TO TAKE A BEARING ON U IMPOSSIBLE TO TAKE A BEA-
  RING ON 3105 - PLS ACKNOWLEDGE THIS XMISION WID A3 ON 3105 GA / UNANSWD  11
KHAQQ DE NRUI RPTING ABVE INFO ON 7500 / NO ANSWR                          12-14
KHAQQ FM ITASCA DO U HR MY SIGS ON 7500 KCS OR 3105 KCS PLS ACKNWLDGE
  WID RECEIPT ON 3105 KCS WID A3 GA / UNANSWD                              15
KHAQQ DE NRUI REPTED ABVE DPE ON 7500 GA 3105 A3 AR / UNANSWD              16-7
KHAQQ FM ITASCA WL U PLS Y OUR SIGS ON 7500 OR 3105 GA WID 3105 A3         18
KHAQQ UNANSWD                                                              19
KHAQQ DE NRUI GA 3105 A3 KCS WID REPORT OUR SIGS                           20-3
KHAQQ DE NRUI GA 3105 KCS WID A3 ES XMIT POSN REPT ES QSA ON OUR SIGS      24-6
ITASCA TO EARHART WE XMITING CONSTANTLY ON 7.5 MEGS DO U HR US KINDLY
  CFM RECEIPT ON 3105 WE ARE STANDING BY, A3/3105                          27
KHAQQ DE NRUI ANS 3105 A3 K / UNANSWD                                      28-9
KHAQQ DE NRUI ANS 3105 KCS WITH REPORT ES POSN, XIXX 7500 / UNANSWD        30-1
EARHART FM ITASCA WL U PLS CUM IN AND ANS ON 3105 WE ARE XMITING CONS-
  TANTLY ON 7500 KCS WE DO NOT HR U ON 3105 PLS ANS 3105 GA / UNANSWD      33
KHAQQ DE NRUI ANS 3105 KCS WID A3 HW OUR SIG QSA? GA / UNANSWD             34-41
LSNIN 3105 / NIL - CRM TUNIN UP T16 FER XMISION TO NMC                     XX 42
NMC V NRUI P AR, 12600 / UNANSWD                                           XX-43
KHAQQ TO ITASCA WE ARE ON THE LINE 157 337 WE WL REPT MSG WE WL REPT N ES S
  THIS ON 6210 KCS WAIT, 3105/A3 S5 (?/KHAQQ XMISION WE ARE RUNNING ON XX LINE
LSNIN 6210 KCS / KHAQQ DE NRUI HRD U OK ON 3105 KCS RX, 7500               44-6
KHAQQ DE NRUI PLS STAY ON 3105 KCS DO NOT HR U ON 6210 MAINTAIN QSO
  ON 3105, 7500 / UNANSWD                                                  47
NIL ON 3105 OR XX 6210 FM KHAQQ / KHAQQ DE NRUI ANS 3105 KCS               48
KHAQQ DE NRUI ANS 3105 KCS A3 / UNANSWD                                    49-53
KHAQQ DE NRUI UR SIGS OK ON 3105 GO AHEAD WITH POSN ON 3105 OR 500
  KCS / UNANSWD                                                            54-0907
LSNIN 3105 AND 6210/500 KCS / NIL                                          08.
KHAQQ DE NRUI ANS 3105 OR 500 UR SIGS OK ON 3105 GA WID POSN, 7500         09-13
LSNIN 6210 ES 3105 - NIL / 500 NIL                                         14
LSNIN 3105, 6210, 500 AND 500 D/F - NIL                                    15-33
NRU12 V NRU1 - PER TO, REF HI FREQUENCY D/F, 7500                          34
KHAQQ DE NRUI GA ON 3105 KCS, 7500 / UNANSWD                               35
NIL FROM KHAQQ / 3105, 6210, 500 OR 500 D/F                                36-41
KHAQQ DE NRUI: XMITING DPE ON ANSWING FREQS / WE CAN HEAR U FINE ON
  3105 PLS GA ON 3105, 7500 / UNANSWD                                      42-6
NIL FROM KHAQQ                                                             49
NIL FROM KHAQQ / 3105, 6210, 500 OR 500 D/F                                52-9
NRU12 V NRU1: - PER TO / GET THE RDO COMPASS WRKING NOW, 7500              1002-2
KHAQQ DE NRUI WE HEARD YOU ON 3105 KCS, ETC - 3500                         03-5
NPM NPU V NRUI V-S ZME / NPM V NRUI INT ZCT KHAQQ INT ZCY KHAQQ AR         08-10
NRUI V NPM ZZA INT FREQ AR / 3105 / R ZOZ AR / DID U HERE KHAQQ AT
  ALL AR / ZZA /R THAT T TO NPU / R (NPM/NRUI — 12600/13380) - PER TO      12
NRUI V NPM AR / NPM V NRUI RK / BNXR3 B NR3 / R AS                         14
TO ENGAGED NPM: XMITING P MSGS, 12600                                      15-26
DC ENGAGED NPM: RECVING P , 12600 / LSNIN 3105 CONSTANTLY, NIL / BG        27-30
BG OFF TO DC//                                                             35
WGEN WGEN DE NRUI NRUI QRU IMI                                             38
WGEN WGEN DE NRUI QRU 600 K                                                39
                        (OVER)
```

Original Bellarts ITASCA radio log 1-2 Jul 37 (Third Page)

```
Form 2614-A - Revised Oct 1935
TREASURY DEPARTMENT
U.S. Coast Guard

RADIO LOG  USCGC ITASCA                    Date  1 July 1937   193

L.G. BELLARTS CRM ON WATCH        ENTRIES    3105 kcs                     TIME

WATCH STARTED AT // SIGNALS HEARD LOGGED ONLY                             1900
VERY WEAK SIGS ON 3105   UNREADABLE AND SEEMED SHIFT ABOUT                1917
NO SIGS ON DURING PERIOD                                                  1945/48
WATCH RELIEVED BY G E THOMPSON RM3C          L.G. BELLARTS CRM            50
NO SIGS ON 3105                                                           2815/18
KHAQQ DE NRUI A A A A NRUI HOWLAND ISLAND A A A A                         2030-33
UNREADABLE FONE SIGS ON OR ABT 3105                                       35/40
SENT WEA TO KHAQQ ON 7500                                                 53-56
KHAQQ DE NRUI A A A A NRUI HOWLAND ISLAND A A A A A                       2100-08
LISTENED FOR KHAQQ ON 3105   ND                                           15-18
SENT WEA 7500                                                             53-56
SENT WEA 3105 / KHAQQ DE NRUI AAAAA NRUI HOWLAND A A A A                  2130-33
UNREADABLE FONE SIGS ON OR ABT 3105                                       41-43
LISTENED FOR KHAQQ - SEVERAL CARRIERS BUT UNREADABLE                      45-48
GAVE WX TO KHAQQ 7500                                                     53-56
KHAQQ DE NRUI AAAAAA NRUI HOWLAND ISLAND AAAAAA                           2200-03
SEVERAL UNREADABLE CARRIERS ON OR ABOUT 3105                              03-07
PLANE UNHEARD                                                             15-18
KHAQQ DE NRUI SENT WEA 7500                                               30-33
KHAQQ DE NRUI AAAAAAA NRUI HOWLAND AAAAAAA                                40-33
LISTENED FER KHAQQ  SEVERAL UNREADABLE FONES BUT CANT MAKE OUT            45-48
GAVE WX TO KHAQQ ON 7500                                                  25-28
KHAQQ DE NRUI AAAAAAA NRUI HOWLAND AAAAAAAA                               30-34
LISTENED FER KHAQQ BUT ND                                                 45-48
SENT WEA TO KHAQQ                                                         55-59
KHAQQ DE NRUI AAAAAA NRUI HOWLAND AAAAAAA                                 2300-03
LISTENED FER KHAQQ BUT ND                                                 15-18
SENT WEA TO KHAQQ ON 7500                                                 25-28
KHAQQ DE NRUI AAAAAAA NRUI HOWLAND AAAAAA                                 30-33
LISTENED FER KHAQQ ON 3105 ND QRM FONES ES UNREADABLE                     45-48
KHAQQ DE NRUI (GAVE HER OUR WEA ON 7500)                                  55-58
KHAQQ DE NRUI AAAAA NRUI HOWLAND AAA PSE QA 3105 NOW K (UNHEARD) (2 July) 0000-04
KHAQQ UNHEARD                                                             15-18
SENT WX TO KHAQQ ON 7500                                                  25-28
KHAQQ DE NRUI AAAAA NRUI HOWLAND AAAAA FER ANS 3105 (unheard)             30-34
KHAQQ UNHRD                                                               45-48
SENT WEA TO KHAQQ                                                         55-58
KHAQQ DE NRUI AAAAAA NRUI HOWLAND AAAAA                                   0100-04
KHAQQ UNHRD                                                               15-18
SENT WEA TO KHAQQ                                                         25-28
KHAQQ DE NRUI AAAAA NRUI HOWLAND AAAAAAAA                                 30-34
KHAQQ UNHEARD                                                             45-48
SENT WEA TO KHAQQ ON 7500                                                 55-58
KHAQQ DE NRUI AAAAA NRUI HOWLAND AAAAA                                    0200-04
WATCH RELD BY LG BELLARTS CRM      G E THOMPSON RM3C                      05
NOTHING HEARD ON 3105 KCS                                                 15-20
SENT WEA ON 7500                                                          28
AAAAAA   AAA HOWLAND NRUI                                                 30-33
ITASCA TO EARHART / fone 3105                                             35
HEARD EARHART PLANE BUT UNREADABLE THRU STATIC                            45-48
SENT WEATHER TO KHAQQ                                                     0300
AAAAAAAA ETC ETC T  KHAQQ HOWLAND DE NRUI                                 04
                         (OVER)
```

ENTRIES	TIME
REPEATED WX ON FONE	0305
NOTHING HEARD FROM EARHART	15-18
SENT WX AT	30
AAAAAAA DE NRUI HOWLED	34
REPEAT WEATHER ON FONE 3105	35
EARHART HEARD FONE WILL LISTEN ON HOUR AND HALF ON 3105 - SEZ SHE	45
BROADCAST WEATHER FONE 3105	0400
REPEATED WEA ON KEY 3105 KCS	02
EARHART UNHEARD	15-18
TRIXST WX ON KEY AND FONE	30-35
HEARD EARHART — PART CLDY	53
SENT WEATHER CODE FONE 3105 KCS	0500
EARHART UNHEARD	12-18
SENT WX AT FONE KEY 3105	30-35
AAAAAAAA NRUI TO 7500 KCS	40
NO HEAR DURING {{SETTING ON 3105 ALL TIMES }}	45/50
SENT WEATHER CODE AND KEY 3105.KCS	0600-05
WANTS BEARING ON 3105 KCS. ON HOUR WILL WHISTLE IN MIC	14
ABOUT TWO HUNDRED MILES OUT APPX WHISTLING NW	15
AAAAAAAA NRUI (7500) (LISTENING THRU ON 3105)	20
AAAA NRUI ——	35
CALLED EARHART ON 3105 VOICE	38
AAAAAAAA NRUI NRUI TO PSE ACKN 3105	41
PSE TAKE BEARING ON US AND REPORT IN HALF HOUR	45
I WILL MAKE NOISE IN MIC ABT 100 MILES OUT	46
AAAAAAAAAA 7500	0705
AAAAAAA 3105	08-18
AAAAA NRUI ETC 7500	12-14
AAAAAAA NRUI 3105	14-18
FONE TO EARHART CANNOT TAKE BEARING ON 3105 VY GOOD PLEASE SEND ON 500 or DU WISH TAKE BEARING ON US GA PSE NO ANSWER'	18
KHAQQ DE NRUI A 5 GA 3105 UNANSWD BELLARTS OFF TO GALVES	19/4
KHAQQ FM ITASCA PLS GA ON 3105 KCS UNANSWD / CRM DO NOW ON THE NRUIS	25
KHAQQ DE NRUI A 5 GA 3105 UNANSWD MAIN D/F 500 KCS	26-9
KHAQQ FM ITASCA PLS Y OUR SIGS ON KEY PSE UNANSWD	30
KHAQQ DE NRUI A-5	31-4
KHAQQ DE NRUI A-5 7500 KCS	35-40
KHAQQ DE NRUI A-5 3105 KCS	41
KHAQQ CLNG ITASCA WE MUST BE ON YOU BUT CANNOT SEE U BUT GAS IS RUNNING LOW BEEN UNABLE TO REACH YOU BY RADIO WE ARE FLYING AT A 1000 FEET	42
KHAQQ DE NRUI R MSG R QSA 5 R A-5 (500 KC 3105) GA UNANSWD	43-5
KHAQQ DE NRUI R MSG R QSA 5 R A-5 (3105)	47-48
KHAQQ FM ITASCA UR MSG OK PLS Y WID AS AS / 3105 / a-s FM NRUI A-5	49-50-57
KHAQQ CLNG ITASCA WE ARE CIRCLING BUT CANNOT HR U GA ON 7500 WID A LNG COUNT EITHER NW OR ON THE SKD TIME ON ½ HOUR (KHAQQ SD A 3)	58
KHAQQ DE NRUI A-3 7500 / GA 3105	
KHAQQ CLNG ITASCA WE RECD UR SIGS BUT UNABLE TO GET A MINIMUM PSE TAKE BEARING ON US AND ANS 3105 WID VOICE / NRUI DE KHAQQ LONG DASHES ON 3105 /	0800
NRUIS de nrui P AR	04
NRUIS GIVING TO DFR TT NO SIGS ON 3105 es IMPOSSIBLE TO MNK ITA /R	
KHAQQ FM ITASCA UR SIGS RECD OK WE ARE UNABLE TO HEAR U TO TAKE A BEARING IT IMPRACTICAL TO TAKE A BEARING ON 3105 UR VOICE HW DO U GET TT GA UNANSWD 3105	05
KHAQQ DE NRUI GA ON 3105 or 500 K , 7500	08

National Archives copy ITASCA radio log 1-2 Jul 37 (Second Page)

```
RADIO LOG    ITASCA                           Date    2 JULY 1937   /93
                         ENTRIES                                    TIME
W L GALTEN, RM3C ON ---- CG32-1
KHAQQ KHAQQ DENRUI NRUI GA 500 WID DASHES ON 500 K/ UNANSWD        0805-8
KHAQQ FM ITASCA DID U GET TT XMISION ON 7.5 MEGS GA ON 500 KCS      80
IT WE MAY BE ABLE TO TAKE A BEARING ON U IMPOSSIBLE TO TAKE A
BEARING ON 3105 PLS ACKNWLDGE THIS XMISION WID A3 ON 3105 GA/
UNANSWD                                                             11
KHAQQ DE NRUI RPTING ABVE INFO ON 7500 KCS ER/ NO ANSW              12-4
KHAQQ FM ITASCA DO U HR MY SIGS ON 7500 KCS OR 3105 KCS PLS ACKNOWLDGE
WID RECEIPT ON 3105 KCS WID A3 GA/ UNANSWD                          15
KHAQQ DE NRUI REPTED ABVE DPE ON 7500 GA 3105 A3 AR/ UNANSWD        16-8
KHAQQ FM ITASCA WL U PLS Y OUR SIGS ON 7500 OR 3105 A3              18
KHAQQ UNANSWD                                                       19
KHAQQ DE NRUI GA 3105 A3 KCS WID REPORT OUR SIGS                    20-3
KHAQQ DE NRUI GA 3105 KCS WID A3 ES XMIT POSN REPT ES QSA ON OUR SIGS25
ITASCA TO EARHART WE XMITING CONSTANTLY ON 7.5 MEGS DO U HR US KINDLY
CFM RECEIPT ON 3105 WE ARE STANDING BY A3/ 3105                     27
KHAQQ DE NRUI ANS 3105 A3 K/ UNANSWD                                28-9
KHAQQ DE NRUI ANS 3105 KCS WITH REPORT ES POSN 7500/ UNANSWD        30-1
EARHART FM ITASCA WL U PLS CUM IN AND ANS ON 3105 WE ARE XMITING CONS-
TANTLY ON 7500 KCS WE DONOT HR U ON 3105 PLS ANS 3105 GA/ UNSWD     33
KHAQQ DE NRUI ANS 3105 KCS WID A3 HW OUR SIG QSA? GA/ UNANSWD       34-41
LSNIN 3105/ NIL CRM TUNIN UP T16 FER XMISION TO NMC                 42
NMC V NRUI P AR.12600/ UNANSWD                                      43
KHAQQ TO ITASCA WE ARE ON THE LINE 157 337 WL REPT MSG WE WL REPT N ES S
THIS ON 6210 KCS WAIT, 3105/ A3 AS (7/ KHAQQ XMISION WE ARE RUNNING ON
LINE LSNIN 6210 KCS/ KHAQQ DE NRUI HRD U OK ON 3105 KCS 7500        44-6
KHAQQ DE NRUI PLS STAY ON 3105 KCS DO NOT HR U ON 6210 MAINTAIN Q50
ON 3105, 7500/ UNANSWD                                              47
NIL ON 3105 OR 6210 FM KHAQQ/ KHAQQ DE NRUI ANS 3105 KCS            48
KHAQQ DE NRUI ANS 3105 KCS A3/ ANANSWD                              49-53
KHAQQ DE NRUI UR SIGS OK ON 3105 GO AHEAD WITH POSN ON 3105 OR 500
KCS/ UNANSWD                                                        54-0907
LSNIN 3105 AND 6210 /500 KCS/NIL                                    08
KHAQQ DE NRUI ANS 3105 OR 500 UR SIGS OK ON 3105 GA WID POSN 7500   09-13
LSNIN 6210 ES 3105- NIL/500 NIL                                     14
LSNIN 3105, 6210, 500 AND 500 D/F- NIL                              15-33
NRU12 V NRUI PER TO REF HI FREQUENCY D/F, 7500                      34
KHAQQ DE NRUI GA ON 3105 KCS, 7500/ UNANSWD                         35
NIL FROM KHAQQ/3105,6210, 500 OR 500 D/F                            36-41
KHAQQ DE NRUI: XMITING DPE ON ANSWING FREQS/ WE CAN HEAR U FINE ON
3105 PLS GA ON 3105, 7500/ UNANSWD                                  42-6
NIL FROM KHAQQ                                                      49
NIL FROM KHAQQ/ 3105, 6310, 500 OR 500 D/F                          52-9
NRU12 V NRUI: PER TO/ GET THE RDO COMPASS WRKING NOW, 7500          1002-2
KHAQQ DE NRUI WE HEARD YOU ON 3105 KCS, ETC - 7500                  03-5
NPM NPU V NRUI V-S ZUE/ NPM V NRUI INT ZCT KHAQQ INT ZCY KHAQQ AR   08-10
NRUI V NPM ZZA INT FREQ AR/ 3105/ R ZOZ AR/ DID U HR KHAQQ AT
ALL AR/ ZZA/ R THAT TO TO NPU/ R (NPM/ NRUI 12600/13380-PER TO      12
NRUI V NPM AR/ NPM V NRUI R K/ B NR3 / R AS                         14
TO ENGAGED NPM: XMITTING P MSGS, 12600                              15-26
DC ENGAGED BON: RECVING P, 12600/ LSNIN 3105 CONSTANTLY, NIL/BG     27-3
BG OFF TO DC                                                        35
WCEN WCEN DE NRUI NRUI QRU IMI (OVER)                               35
WCEN WCEN DE NRUI QRU IMI 600 K
```

National Archives copy ITASCA radio log 1-2 Jul 37 (Third Page)

```
Form 2614 A—Revised Oct. 1931
TREASURY DEPARTMENT
U. S. Coast Guard
```

RADIO LOG ___ ITASCA ___   Date 2 JULY 1937 ___ 193_

ALL OF THE "TIME" IS COMPUTED ENTERZONE PLUS 10 1/2. | TIME

F CIPRIANI ON HOWLAND ISLAND STANDING BY TO TAKE BEARINGS ON 3105 KCS ON EARHART PLANE. ASSUMED WATCH AT ----	2200
WEAK FONE ON 3105 KCS. UNREADEABLE	2345
3 JULY 1937	
WEAK FONE ON 3105 (I AM USING A LONG VERTICAL ANTENNA FOR RECEPTION OF SIGNALS ONLY) UNABLE TO GET BEARINGS	0015
ITASCA TESTING WITH NMC ON FONE 3105	0215
ITASCA GIVING WEATHER ON FONE TO KHAQQ BLIND 3105	0630
ITASCA GIVING WEATHER ON FONE TO KHAQQ BLIND 3105	0700
ITASCA GIVING WEATHER ON FONE TO KHAQQ BLIND 3105	0715
PICKED UP EARHART (USING LONG ANTENNA, S3, HARDLY ANY CARRIER, SEEMED OVERMODULATED, SWITCHED OVER TO LOOP FOR BEARING, S1 TO 0. SHE STOPPED TRANSMISSION) BEARING NIL 3105	0717
WORKED ITASCA (REQUESTING BEARING ON PLANE) NRUI V NRUI2 R	0725
ITASCA SENDING BLIND TO KHAQQ	0735
(AM USING THE D/F AND RECEIVING SET SPARINGLY DUE TO HEAVY DRAINAGE ON BATTERIES) (THE BATTERIES ARE OF LOW AM-HOUR CAPACITY) EARHART ON THE AIR, S4, "GIVE ME A BEARING". EARHART DID NOT TEST FOR BEARING. HER TRANSMISSION TOO SHORT FOR BEARING, STATIC X5, HER CARRIER IS COMPLETELY MODULATED. COULD NOT GET A BEARING DUE TO ABOVE REASONS. 3105	0747
ITASCA SENDING "A" TO KHAQQ 3105	0800
ITASCA SENDING "A" TO KHAQQ 3105	0815
ITASCA CALLING KHAQQ ON FONE 3105	0820
ITASCA CALLING KHAQQ ON FONE 3105	0830
KHAQQ DE NRUI NRUI MSG MSG R R AAAAAA ETC 3105	0845
BATTERIES WEAK	
VOICE ON 3105. CAME IN AT END OF TRANSMISSION. 3105	0859
ITASCA CALLING EARHART TO ANSWER ON 500 KCS STEADLY TILL ----	0920
RECEIVED INFORMATION THAT ITASCA BELIEVE EARHART DOWN. LANDING PARTY RECALLED BACK TO VESSEL.	0926
ALL BATTERIES ON THE ISLAND ARE DISCHARGED. COMMENCED TO CHARGE THEM.	1000
4 JULY 1937	
CHARGING BATTERIES ALL DAY	

F CIPRIANI RM2C

TRANSCRIPT OF THE LOG KEPT ON HOWLAND ISLAND

(OVER)

Howland Island radio log RM2 Cipriani 2 Jul 37
(National Archives copy)

Appendix VIII -

The Morgenthau Transcript and Related Documents

UNITED AIR SERVICES, Ltd.
BURBANK, CALIFORNIA

EXECUTIVE OFFICES

April 26th 1938.

Mrs. Franklin D. Roosevelt,
The White House,
Washington, D.C.

Dear Mrs. Roosevelt:

      The writer acted as technical advisor to Miss Earhart and made the flight with her to Honolulu which was climaxed by her accident on the take-off.

      I received a letter from Miss Jacqueline Cochran the other day requesting that I answer a great many questions regarding my opinion and actual messages that were received and tied together with the times in relation to the take-off.

      I have often wanted to make up a complete report in detail on this situation in order to determine whether or not a search would be practical, even at this late date. In order to do so would like to have a copy of the official report of the "Itasca" which is on file at the Coast Guard Headquarters in Washington. I saw this report in San Francisco at the time Miss Earhart was first reported missing but made no notes and if you could arrange to have a copy sent to me I would in turn furnish you a detailed report and theories for your own personal use - but not for publication.

      I have attempted to draw up this information ever since last August but just this morning I was advised by Mr. F. K. Johnson of the U.S.Coast Guard, San Francisco, that the official report was on file in Washington and could not be released except through certain channels.

Letter from Paul Mantz to Eleanor Roosevelt of April 26, 1938 – Page 1

Mrs. F. D. Roosevelt   -2-

      I feel that all of us who were so close to Miss Earhart should at least make a certain amount of effort along the lines I know she would follow if the situation were reversed.

      Would deeply appreciate it if you could have the above information forwarded and in turn you will receive a report which I am sure will answer any question of doubt that may be in your mind.

                        Yours very truly,
                        UNITED AIR SERVICES, LTD.

APM:S   A. Paul Mantz, President

Letter from Paul Mantz to Eleanor Roosevelt of April 26, 1938 – Page 2

/100

6/27
Sent to Mr. Morgenthau - "Can he do this or what do I say? E.R."
Paul Mantz, Burbank, Calif.
wants Coast Guard data, which was originally turned over to him and he did not copy.

Administrative note dated 6-7-38

THE WHITE HOUSE
WASHINGTON

May 10, 1938

Dear Henry:

    A little while ago Floyd ~~Oldrum~~ Odlum and his wife, Jacqueline Cochran, were at the White House when she received the Harmon Trophy for aviation. She told me they all felt that not enough search had been made amongst certain islands where Amelia Earhart might be. I told her to send me a memo on the Islands and the reasons why they felt this, and I would transmit it to you and to the Navy Department at once. Now comes this letter which is evidently inspired by Miss Cochran. I do not know whether you can send the man these records, but, in any case, I am sending you the letter and let me know whatever your decision may be.

Affectionately,

Note to Henry Morgenthau from Eleanor Roosevelt of May 10, 1938

-15-

Here's the confidential information: "...and a minimum of $70,000 a week is needed." He's got the whole thing.

Yes, yes.

Mr. Secretary, may I suggest to you the possible advantage of having a representative of Social Security at your conference?

I don't think the President would stand for it. I don't think that he'd stand for it, Miss Lonigan. I don't want to get excited, but Social Security and the Hopkins organization sit here and do nothing.

Social Security can't, but they have the facts.

But they don't talk. They should be vocal.

Well, I agree with that.

I mean how can a person have anything to do with an organization called "social" and sit here and see people - and see people starve?

They couldn't spend a dollar.

But they could be vocal. I mean there's such a thing as bringing it to somebody's attention, and it just - it just turns my stomach to hear Henry Wallace want a hundred million dollars to have the people grow less wheat and then - and with people not getting enough to eat. Now, there's just something cock-eyed, crackpot, about this administration. I mean it just goes against all decency and human understanding that they should be trying to find ways and means to grow less.

Yes.

And there's people going hungry in America, all over America. Now, there's just something - the combination of Wallace and Hopkins refusing to do any direct relief - just something ungodly about it.

And draw the money from a mass consumers' tax, which is .....

Morgenthau Transcript, telephone conversation May 13, 1938 Pg. 411

-16-

And I'm not going to keep quiet.

(On White House phone) Oh, hello. - Oh, thanks. Hello, Tommy (Malvina Scheider). How are you? This letter that Mrs. Roosevelt wrote me about trying to get the report on Amelia Earhart. Now, I've been given a verbal report. If we're going to release this, it's just going to smear the whole reputation of Amelia Earhart, and my ..... - Yes, but I mean if we give it to this one man we've got to make it public; we can't let one man see it. And if we ever release the report of the Itasca on Amelia Earhart, any reputation she's got is gone, because - and I'd like to - I'd really like to return this to you.

(Continuing) Now, I know what Navy did, I know what the Itasca did, and I know how Amelia Earhart absolutely disregarded all orders, and if we ever release this thing, goodbye Amelia Earhart's reputation. Now, really - because if we give the access to one, we have to give it to all. And my advice is that - and if the President ever heard that somebody questioned that the Navy hadn't made the proper search, after what those boys went through - I think they searched, as I remember it, 50,000 square miles, and every one of those planes was out, and the boys just burnt themselves out physically and every other way searching for her. And if - I mean I think he'd get terribly angry if somebody - because they just went the limit, and so did the Coast Guard. And we have the report of all those wireless messages and everything else, what that woman - happened to her the last few minutes. I hope I've just got to never make it public, I mean. - O.K. - Well, still if she wants it, I'll tell her - I mean what happened. It isn't a very nice story. - Well, yes. There isn't anything additional to something like that. You think up a good one. - Thank you. (Conversation ends)

(To Chauncey) Just send it back.

Sure.

I mean we tried - people want us to search again

Morgenthau Transcript, telephone conversation May 13, 1938 Pg. 412

those islands, after what we have gone through. You (Gibbons) know the story, don't you?

We have evidence that the thing is all over, sure. Terrible. It would be awful to make it public.

Well, the only thing that out of this - I want you (Lonigan) to check up with Social Security. Archie, give this (photostat of WPA figures) to her and let her check, and Ed, you check the legislation, will you please?

Yes. Here's an unsigned memorandum.

Excuse me?

There is what Dan said about the law.

And you two can be excused, please - Miss Lonigan and Mr. Foley.

Mr. Secretary, there is one factor you might wish. Yesterday Mr. Gill called it to you - the average weekly payment for direct relief was $21 a week, and I checked on what F.E.R.A. was paying in 1935, and they had full control of general relief and they paid $7.12 a week. That's the discrepancy.

Is that a family or person?

Everything is the family.

That's a United States average.

No, I think it's the city of Cleveland.

Well, it's four and a half now.

They mean the maximum you'd need for full programs.

I know. I mean - well, as I recall, it was around $28 a month.

A dollar a day.

Those figures are all drilled in my brain. I never forget them. I don't know what his object was.

(Lonigan and Foley leave)

Morgenthau Transcript, telephone conversation May 13, 1938 Pg. 413

May 14, 1938

My dear Mr. Mantz:

     I have made inquiries about the search which was made for Amelia Earhart and both the President and I are satisfied from the information which we have received that everything possible was done. We are sure that a very thorough search was made.

         Very sincerely yours,

Mr. A. Paul Mantz
United Air Services, Ltd.
Burbank, California

S:DD

Carbon of letter to Paul Mantz from Eleanor Roosevelt May 14, 1938

THE WHITE HOUSE
WASHINGTON

Mr. Morgenthau says that he can't give out any more information than was given to the papers at the time of the search of Amelia Earhart

It seems they have confidential information which would completely ruin the reputation of Amelia and which he will tell you personally some time when you wish to hear it.

He suggests writing this man and telling him that the President is satisfied from his information, and you are too, that everything possible was done.

Undated note on White House stationery

THE SECRETARY OF THE TREASURY
WASHINGTON

Dear Eleanor:

We have found it possible to send to Mr. A. Paul Mantz a copy of the log of the ITASCA, which I think will supply him with all the data he asked for in his letter of June 21st.

Sincerely,

Henry

Mrs. Eleanor Roosevelt,
The White House.

Note to Eleanor Roosevelt from Morgenthau 7-5-38

July 5, 1938

Dear Henry:

Thank you very much for your note, informing me that you found it possible to send a copy of the log of the ITASCA to Mr. Paul Mantz, whose letter I had referred to you.

Very sincerely yours,

Morgenthau

70
MORGENTHAU

Note from Eleanor Roosevelt to Morgenthau 7-5-38

```
                                                          OP-001
                                                            6071
                                                              61
```

21 July, 1938.

Mr. A. Paul Mantz,
    Burbank,
        California.

Sir:

      Your letter of June 21, 1938, addressed to Mrs. Eleanor Roosevelt, The White House, Washington, D. C., has been referred to this office for consideration.

      In reply thereto, I am pleased to forward herewith, copies of the transcripts of the log of the Coast Guard Cutter ITASCA, which pertain to the Amelia Earhart flight, together with copies of messages received and sent incident to this flight.

                  Very truly yours,

                          R. R. WAESCHE,
                          Rear Admiral, U. S. Coast Guard,
                          Commandant.

Inclosures:

Copy to: Mrs. Franklin D. Roosevelt,
         The White House,
         Washington, D. C.

Letter from RADM Wesche to Paul Mantz of July 21, 1938

Appendix IX -

Pros and Cons - Was Irene Bolam Amelia Earhart?

What follows is my attempt to elucidate both the pro and con arguments regarding whether or not Irene Bolam was actually Amelia Earhart. I undertook this purely as intellectual effort and came up with every argument I could think of, without really knowing how it would turn out. Interestingly, when I finished, the arguments in favor of Irene Bolam having been Amelia Earhart are a little stronger. Was she?

Arguments that Irene Bolam was Amelia Earhart:

- ❑ The extremely close physical resemblance between the two as observed in Tod Swindell's study. The resemblance appears to be too close to be coincidental. The preliminary report of the forensic scientists was that they felt that the coincidence of bone structure was exact.

- ❑ The full physique and ample bosom in the Gervais photo may have been the result of hormonal changes and/or illness Earhart experienced as the result of stress and poor diet in her captivity.

- ❑ Bolam's own admission to Joe Gervais over the phone that she had "been a public figure, had a career in aviation, and had left it all behind." Historically, Irene Bolam was never a public figure and did not have an actual career in aviation.

- ❑ Bolam moved extensively in old time aviation circles and received hugely deferential treatment. This is at odds with Irene Bolam's reported miniscule aviation involvement (brief lessons and flying time in 1933. Also the fact that she claimed to be a Ninety-Nine and a Zonta, but her name was in neither organization's roster.

- ❑ Close friends of Bolam's noted her strong physical resemblance to Earhart. They also noted she made many peculiar remarks that inferred she was Earhart, and some were convinced that Bolam was indeed Earhart. Even Bolam's own doctor felt she might be Amelia Earhart.

- Bolam was privately given an award by NASA in the late 1960s. This is again at odds with Bolam's recorded microscopic aviation experience.

- Bolam's insistence on never being officially identified alive or dead. This first surfaced during her lawsuit against Joe Gervais and Joe Klaas. She actually dismissed her claim in the lawsuit and gave up a sizeable settlement to avoid giving her fingerprints to the court.

- Bolam's insistence to Joe Gervais that she could never discuss Amelia Earhart "in this country". She made an appointment to meet Gervais in Canada, which she later broke. Bolam's friends were also very skittish about discussing Amelia Earhart with her.

- Bolam was recognized as Earhart by both Joe Gervais and Robert Myers. The two men did not know or have any connection with each other

- Bolam's brother-in-law and sister-in-law both stated that they felt Bolam was not who she seemed to be.

- Bolam was influential, very well connected politically and wealthy, owning several homes. The real Irene Bolam was dismissed as a typical middle class housewife.

- Bolam's husband's given business affiliation was reported by Joe Gervais to be a dummy corporation. Bolam's husband had an intelligence background and was said to be an agent for the British MI6 agency. These circumstances seem a bit esoteric for a typical middle class housewife with a slight aviation history.

Arguments that Irene Bolam was not Amelia Earhart:

- The Swindell Photo Study has not been finalized and is not conclusive.

- The resemblance between Irene Bolam and Amelia Earhart has been overplayed. In addition, in the 1965 Gervais photo, Bolam was shown with a very ample physique and full bosom, characteristics Earhart had not had in her lifetime.

- Bolam had a right to prevent her identification to protect her privacy. Therefore her various actions and behaviors may have been highly eccentric but are not necessarily suspect.

- Joe Gervais and Robert Myers were mistaken in their recognition of Bolam as Earhart.

- Many of Bolam's relatives have strongly insisted that Bolam couldn't have been Earhart because they had known her for many years and her life had been low-key and unremarkable.

- Although she had a miniscule aviation record, Bolam was one of the few surviving participants in early aviation and thus was accorded deference by old-time flyer organizations and the establishment alike.

- There is no hard evidence that Amelia Earhart ever assumed the identity of Irene Bolam.

- Irene Bolam may have actually been an "agent of distraction" for those who wanted to divert attention of researchers away from Earhart's real fate.

- Irene Bolam may have actually been a fabricator herself, like Msgr. Kelly, and enjoyed confusing friends and researchers.

Appendix X -

History of Howland Island

Since Howland Island is of such central importance in the disappearance of Amelia Earhart, the following is presented for the reader's information.

Howland Island is situated 1650 miles southwest of Honolulu, Hawaii and just 48 miles north of the equator. It is also northwest of the Phoenix Island Group.

Just over a mile and a half long, Howland is a half mile wide. Although there are elevations on the island of approximately 18 to 20 feet, the island is mostly quite flat and little more than a foot or two above sea level.

Quite warm, but not uncomfortably so, the island only experiences noticeable heat under the midday sun. Due to the flat nature of the island, no clouds form to speak of over the island, and as a result the climate is relatively dry. The wind blows frequently if not incessantly. And of note to the Earhart researcher, the wind blows from the southeast in the summer. That would have meant headwinds for the Amelia Earhart's flight from Lae, New Guinea to Howland Island.

Some slight archaeological sites on the island indicate habitation in the distant past by Polynesians, although Howland had been long uninhabited when in 1822 Captain George Worth landed there in the Nantucket whaler *Oeno*, giving the island its first name, Worth Island.

Over the next twenty years several other vessels landed on the remote island, and finally, in 1842, Captain George Netcher landed on the island in the whaler *Isabella*. Netcher promptly renamed the island after the lookout on his ship, who had sited it.

---

[61] Kansas City Journal-Post, March 14, 1937, pg. 9B.

Beginning in 1857 Howland Island was periodically worked for the bird guano which was found on it huge amounts, and in 1858, The United States Guano Company filed a claim on the island for guano mining.

Between 1886 and 1891, a major occupation occurred on the island, when John T. Arundel & Co. moved 100 laborers from the Niue and Cook Island group to mine guano.

After 1891, the island remained unoccupied until January 1935, when William T. Miller of the U.S. Bureau of Air Commerce secretly brought an expedition of colonists from Honolulu, aboard the USCG Cutter ITASCA, to Howland, Jarvis and Baker Islands.[61] By March 1935, Howland had been colonized, and early the following year, in February 1936, after all the islands had been colonized, the U.S. quietly annexed them.

In November 1936, President Roosevelt ordered the construction of an airstrip on Howland Island to support the world flight of Amelia Earhart. Following the failed world flight, the United States government erected a monument, Amelia Earhart Light, on Howland Island in memory of Amelia Earhart in 1938. The island remained inhabited by colonists until late 1941.

On their way back to Japan from Pearl Harbor on December 7, 1941, Japanese forces brutally attacked Howland Island, destroying all facilities on the island and killing all but two colonists. The surviving colonists were evacuated on January 15, 1942.

In recent years, the Amelia Earhart Light was rebuilt on the same site, still without a light, where it stands a lonely vigil over the empty Pacific to this day.

Appendix XI -

Selected Website List

Below is a selected listing of websites which I visited during my research and which would be of interest to the reader desiring further information.

http://www.tighar.org	TIGHAR website
http://www.cnmi.net/	Commonwealth of Northern Marianas Islands
http://www.cnmi-guide.com/history/ww2/amelia/	CNMI Guide
http://www.guampdn.com/news	Guam Pacific Daily News
http://www.mvariety.com	Marianas Variety News
http://www.nara.gov	U.S. National Archives
http://www.history.navy.mil/faqs/faq3-1.htm	U.S. Naval Historical Center FAQ Page
http://www.ellensplace.net/eae_intr.html	Ellens Place – Earhart biography site
www.ameliaearhartmuseum.org	Amelia Earhart Birthplace Museum
www.ninety-nines.org/earhart.html	Ninety-Nines website
http://www.earhart.org	Earhart Eyewitnesses Web Page (Thomas Devine)
http://www.ameliaearhart.com/home.php	Official Amelia Earhart Website (Earhart Estate)

http://ameliaearhart.org	"Amelia Earhart Survived" - Rollin Reineck's website
http://www.electranewbritain.com	David Billings' website
http://en.wikipedia.org/wiki/Amelia_Earhart	Wikipedia article on Amelia Earhart
http://www.womeninaviation.com/amelia.html	Women in Aviation
http://www.infoplease.com/ipa/A0192891.html	InfoPlease website
http://janeresture.com/howland/	History of Howland Island

Appendix XII -

Photographic Overlays by David Deal

[I was lucky enough to locate and get permission to use the following photographic overlays shortly before publication of this book. What follows are comments by David Deal, creator of the overlays. They are herewith presented to the reader as possible new evidence in the mystery of Amelia Earhart. –The author]

I first became aware that there was a possible cover-up or mystery involving the aviation heroine of my youth, Amelia Earhart, when I read Joe Klass and Joe Gervais' book in about 1971 "Amelia Earhart Lives." I was enthralled with the research that Joe Gervais did. I lent out the book and never got it back, so over the years, I asked many a used book seller to help me find another copy to no avail. I remembered most of the book's details fairly well, because I have that kind of mind. I was greatly relieved when the book was re-released in 2000.

Shortly thereafter I got my copy, and was back in business. I spoke with Joe Gervais several times, and had an ongoing telephone communication with him ever since. My interest was and remains intense on this subject.

But the issue of Irene Bolam is another matter. Here, in this arena, I could produce evidence that would be striking.

Since I am an artist who works a great deal on the computer nowadays, and since I was able to download many pictures of Amelia and Irene Bolam that I had not been aware of before, it was only natural for me to start looking for similarities between these supposedly different women. So I began several years ago to make overlay comparisons. This is very easily accomplished on the omputer using the Photoshop software that I use in my work.

Here is how it is done.

First I must explain, this technique is not computer "morphing," which some people first think when they see my overlays. With "morphing" one can make a monkey turn into a man or vice versa. The technique I use is nothing of the kind. I take an unretouched photograph of Amelia Earhart that appears nearly the same angle as one of the Irene Bolam photos that have circulated. Then I make an overlay and align the faces using teeth, eye orbits, by making the overlay transparent, so that I can see through.

I align both faces this way, then make the overlay opaque again, so that I cannot see the underlying photo.

In small progressive steps, and very smoothly I begin to change the overlay's intensity or transparency using the program's slider control. I take a screen shot of the 100% photo first then reduce transparency to 80%, at this point I begin to see through the overlay. I take another screen shot. I continue this process 10-15% at a time, and each step reveals a little more of the underlying photograph. Finally the underlying base photo is completely viewed at 100% intensity.

There is no manipulation involved in this process. What you see, is one face dissolving into another face. Even though there are 30-35 years in between these photos, the unchangeable structure of bone (and sometimes teeth if they are real), and muscle is clearly seen.

Most people are acutely aware of the face and can recognize when an artist makes a bad picture, so it is this universal familiarity with the human face that enables even the non-artist to see minute details correctly. You must be the judge. I can only offer the evidence.

As for me, personally, I cannot see how these photographs can fail to convince a reasonable person that we have been subjected to a very mysterious case of Legerdemain.

David Allen Deal
January 22, 2005

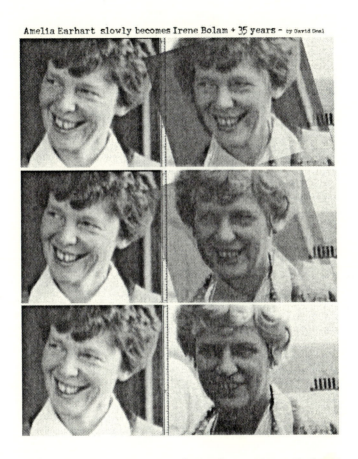

David Deal Overlay: Amelia Earhart to Irene Bolam

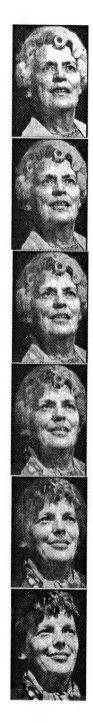

David Deal Overlay: Irene Bolam to Amelia Earhart

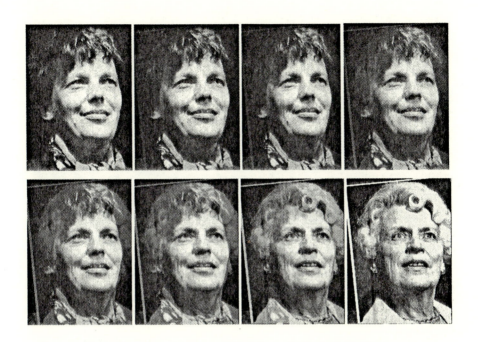

David Deal Overlay: Amelia Earhart to Irene Bolam

Bibliography

Books

ANDREW, Christopher, "For the President's Eyes Only: Secret Intelligence and the American Presidency from Washington to Bush", NY: Harper Collins Publishers, 1995

BRENNAN, T.C., "Witness to the Execution," CO: Renaissance House Publishers, 1988

BRIAND, Paul, "Daughter of the Sky," NY: Duell, Sloan and Pearce, 1960

BRINK, Randall, "Lost Star: The Search for Amelia Earhart," NY: W. W. Norton Co., 1993

CHAPMAN, Sally Putnam, "Whistled Like a Bird," NY: Warner Books, 1997

COCHRAN, Jacqueline, "The Stars at Noon", Boston: Little, Brown and Co., 1954

DAVIDSON, Joe, "Amelia Earhart Returns from Saipan," Bloomington, IN: Unlimited Publishing, 2002

DEVINE, Thomas, with Richard D. Daley, "Eyewitness: The Amelia Earhart Incident," Frederick, CO: Renaissance House, 1987

DEVINE, Thomas, with Mike Campbell, "With Our Own Eyes: Eyewitnesses to the Final Days of Amelia Earhart," Lancaster, OH: Lucky Press, 2002

DONAHUE, James A., "The Earhart Disappearance: The British Connection," Terra Haute, IN: Sunshine House, 1987

DWIGGINS, Don, "Hollywood Pilot: The Biography of Paul Mantz", Garden City, NY: Doubleday Co., Inc., 1967

EARHART, Amelia, "Last Flight," NY: Harcourt, Brace & Co., 1937

GOERNER, Fred, "The Search for Amelia Earhart," NY: Doubleday 1966

GOLDSTEIN, Donald M. and DILLON, Katherine V., "Amelia: A Life of the Aviation Legend" Washington, DC: Brassey's, 1997

KAHN, David, "The Code Breakers", NY: MacMillan, 1967, 9th Ed.

KLAAS, Joe, "Amelia Earhart Lives," NY: McGraw Hill, 1970

KNAGGS, Oliver, "Amelia Earhart: Her Last Flight," Capetown, South Africa: Timmins Publishers, 1983

LONG, Elgen and Marie, "Amelia Earhart—The Mystery Solved", NY: Simon and Schuster, 1999.

MORRISSEY, Muriel, "Courage is the Price," Wichita, KS: McCormick-Armstrong Publishing Division, 1963

MORRISSEY, Muriel, and Osborne, Carol L., "Amelia, My Courageous Sister," Santa Clara, CA: Osborne Publisher, 1987

MYERS, Robert H., "Stand By to Die: The Disappearance, Rescue and Return of Amelia Earhart," Grove, CA: Lighthouse Writers' Guild, 1985

PERSICO, Joseph, "Roosevelt's Secret War," NY: Random House, 2001

PUTNAM, G. P. Jr., "Soaring Wings," NY: Harcourt, Brace & Co., 1939

REINECK, Rollin C., "Amelia Earhart Survived," CA: The Paragon Agency, 2003

ROESSLER, Walter and GOMEZ, Leo, "Amelia Earhart—Case Closed?", Hummelstown, PA: Aviation Publishers, 1996

WILSON, Donald D., "Amelia Earhart: Lost Legend," Webster, NY: Enigma Press, 1994

Articles:

"Amelia Earhart: A postal Portrait," M a rjory J, Sente, Scott's Monthly Stamp Journal, November 1977, Pg. 10-15, Vol. 58, No. 11.

"65 years later, the mystery of Amelia Earhart continues," Ron Staten, AP, athens.com/stories/070202/new_29929702005.html.

"A Woman Hops the Atlantic," Literary Digest, June 30, 1928

BRUDER, Jerry, "The Enduring Mystery of Amelia Earhart," American History Illustrated, May 1987

EARHART, Amelia, "Crossing the Atlantic," American Magazine, August 1932

DiTOMMASO, Lois; Emmons, Sue and Kenyon, Donna, "Did Amelia Die or Was She Irene Bolam?", Woodbridge NJ News Tribune, October 18-29, 1982

GILLESPIE, Richard, "The Mystery of Amelia Earhart," Life Magazine, April 1992

JENNINGS, Dean S., "Is Amelia Earhart Still Alive?" Popular Aviation, December 1939 and January 1940

RILEY, John P., "The Earhart Tragedy: Old Mystery, New Hypothesis," John P. Riley, Naval History Magazine, August 2000

SOTHERN, Robert, "Trapping the Amelia Earhart Extortion Ghouls," True Detective Mysteries, January 1940

"This Boston", The Bostonian, July 1928

Newspapers:

Appleton Post-Crescent
    June 20, 1932; May 4, 1939
Bismarck Tribune, May 8, 1935
Charleston Daily Mail
    July 3, 1937; July 5, 1937; July 9, 1937
Chillicothe Constitution-Tribune ; July 16, 1937
    July 6, 1937; July 12, 1937; July 14, 1937
Chronicle-Telegram
    February 18, 1937; July 6, 1937
Clearfield Progress, July 14, 1937
Coshocton Tribune
    February 16, 1937; July 25, 1937
Edwardsville Intelligencer, 7-9-37
Hammond Times (IL)
    June 8, 1937; July 14, 1937
Helena Daily Independent
    February 26, 1937; July 7, 1937; July 9, 1937
Indiana Evening Gazette, July 13, 1937
Iowa City Press-Citizen, May 20, 1939
Kansas City Journal-Post, March 14, 1937
Lima News

July 4, 1937; July 10, 1937; July 11, 1937
May 13,1939
Mansfield News-Journal, July 17, 1937
Monessen Daily Independent
July3, 1937; July 15, 1937
Nevada State Journal
April 20, 1935; December 13, 1937; April 19, 1939
May 21,1939
Newark Advocate
June 12, 1931; July 5,1937; August 5,1937
New York Times, 9-7-35
New Advocate, August 5, 1937
Reno Evening Gazette
November 5, 1929; July 2, 1937; July 8, 1937
December 16, 1937
Sheboygan Press
June 12, 1931; July 6, 1937; December 21, 1937
Woodbridge NJ News Tribune, October 18-29,1982

Other:

HOOVEN, Frederick, 1982 research paper on the disappearance of Amelia Earhart, Smithsonian National Air & Space Museum Library, Washington, DC.

BRIGHT, Ron, and CAMPBELL, Mike, "The Forrestal Incident," unpublished monograph, 2002

Index

A
AKAGI 67, 68
Amaron, Billamon 133
Amelia Earhart Discussion Group xiii, 45, 67, 151, 198, 207
Amelia Earhart Society 151, 153, 190, 255
Angwin, Donal 183
ATHENE, MV 169, 170, 171, 232, 233

B
Bay Farm Airport (See "Oakland Airport") 71
Bellarts, David xiii, 53, 57, 59, 75, 160
Bicycle Lake 18, 19, 20, 21, 22, 41, 42, 43, 46, 259, 276
Billings, David 182, 183, 304
Black, Richard 76, 213, 224, 245, 247, 248
Blas, Nieva Cabrera 109, 134
Bolam, Guy 24, 27, 30, 31, 150, 153, 248, 249, 250, 251, 253
Bolam, Irene xiii, 22, 23, 24, 25, 26, 27, 30, 31, 32, 70, 71, 77, 149, 151, 152, 153, 154, 157, 173, 192, 193, 194, 195, 203, 205, 207, 208, 214, 243, 249, 250, 252, 253, 254, 298, 299, 300, 305, 307, 308, 309
Bolam, John 27, 249
Bowman, Hong xiii
Bowman, Sydney S. 33
Bowman, Thomas xiii
Brennan, T.C. "Buddy" 106
Bright, Ron Burks, Billy 110, 138, 229

Bunn, Jennings 184

C
Campbell, Mike 86, 96
Carrington, G. C. 158
CENTRAL AMERICA, SS 144
Cervone, Michelle xiii
Chapman, Sally Putnam 36, 173, 256, 257
Cipriani, RM2 Frank 53, 58
Coutts, Alex 198

D
Davidson, Joe Kothera 15, 16
Dawes, Diana 152, 207
Deal, David xiii, 192, 305, 307, 308, 309
DeCarie, Margot 16, 200, 225, 229, 233
Devine, Thomas xiv, 13, 14, 86, 88, 92, 93, 139, 140, 190, 198, 201, 237, 238, 239, 251, 258
Dinger, Robert xiv, 13, 249
Donahue, James R. xiv, 11, 76, 80, 134

E
Earhart, Amelia v, xiii, xiv, xv, xvii, 1, 4, 7, 9, 13, 15, 18, 19, 20, 21, 22, 23, 24, 25, 27, 28, 29, 30, 31, 33, 34, 35, 37, 38, 40, 41, 42, 43, 45, 46, 48, 49, 50, 59, 62, 63, 65, 67, 70, 71, 73, 77, 78, 80, 82, 83, 84, 86, 87, 88, 89, 90, 91, 92, 94, 97, 99, 100, 101, 102, 103, 104, 106, 116, 119, 126, 127, 129, 130, 131, 135, 137, 138, 139, 141, 142, 144, 146, 147, 148, 149, 150, 151, 152, 153, 154, 155, 157, 158, 160, 162, 164, 167, 168, 169, 171, 173, 174,

175, 176, 177, 180, 181, 182,
183, 185, 189, 192, 193, 194,
196, 197, 198, 199, 201, 203,
204, 206, 207, 208, 211, 212,
216, 217, 219, 221, 225, 226,
228, 229, 230, 231, 233, 234,
239, 240, 242, 243, 244, 248,
250, 251, 252, 254, 255, 256,
257, 258, 260, 265, 266, 267,
268, 274, 275, 276, 298, 299,
300, 301, 302, 303, 304, 305,
307, 308, 309
Elliott, Robert 44, 83

F

Federal Bureau of Investigation
76, 77, 84, 160, 161, 167, 187,
223, 226, 252
Firmosa, Fujie 67
Florence Matonis, Ken 17
Forrestal, James 92, 94, 237

G

Galten, RM3 W.L. 53
Garapan Jail 106, 107, 108, 136
Gardner Island (See "Nikumaroro
Island") 51, 147, 254
Gentry, Viola 23, 24, 25, 26, 220,
250
Gervais, Joe xiv, 13, 20, 22, 23,
24, 26, 28, 29, 35, 36, 41, 42,
44, 46, 67, 70, 134, 136, 149,
152, 153, 154, 156, 172, 173,
180, 192, 193, 194, 195, 203,
214, 249, 250, 251, 252, 259,
298, 299, 300, 305
Gilbert Islands 51, 52, 177, 225,
228
Goerner, Fred 13, 14, 15, 30, 50,
95, 96, 135, 139, 151, 159, 202,
238, 245, 249, 250, 255

Gordon, Louis 1
Gradt, William 202, 205
Greene, Wallace 89, 238

H

Hawaii Clipper 179, 180, 233, 234
Heine, Carl 99, 101, 107
Henson, Everett Jr. 110, 138
Hill, Charles 179, 180, 232, 233,
234
Hooven, Frederick 167
Howland Island xiii, 4, 5, 10, 29,
31, 49, 50, 51, 53, 55, 57, 58, 59,
60, 61, 62, 63, 64, 65, 66, 75,
81, 82, 83, 142, 143, 144, 145,
147, 158, 159, 168, 177, 182,
183, 186, 189, 197, 225, 230,
277, 285, 301, 302
Huxford, George 158

I

Imiej Island 133
In Search of Amelia LLC 145
ITASCA, USCGC 119, 120, 121,
128

J

Jack, Lotan 133
Jaluit 9, 67, 68, 99, 100, 101, 102,
103, 107, 133, 134, 159, 190,
198, 200, 205, 226, 227, 229,
233, 237, 247, 265, 266, 267
Johnson, Clarence L. "Kelly" 186

K

Kammerer, Mike 145
KAMOI 68, 227, 247
KAMUI 68
Klaas, Joe xiv, 13, 20, 27, 41, 68,
70, 77, 149, 192, 211, 242, 250,
251, 252, 299
Knaggs, Oliver 103

KOSHU 68, 133, 190, 226, 227, 247
KOSYU 68, 69
Kothera, Donald 15, 17, 200
Kwajelein 107, 134, 137, 158, 159, 212, 226, 236, 239, 240

L
Lockheed Electra Model 10E 80, 185, 186
Lockheed Electra Model 12A 19, 41, 274
Loomis, Vincent 50, 51
Luke Field 4, 10, 44, 46, 48, 71, 84, 141, 142, 186, 198, 225

M
Maleolap Atoll 85, 100
Mandel, Dr. Alex xiii, 67, 68, 86
Manning, Harry 143, 168, 224, 225
Mantz, Paul 4, 15, 18, 21, 22, 46, 62, 127, 128, 129, 130, 131, 142, 186, 197, 223, 224, 258, 274, 275, 276, 286, 287, 293, 297
Marianas History Museum 184
Mayazo, Tomaki 133
McMenamy, Walter 61, 83, 186, 187, 226
Messages in a bottle 227
Mili Atoll 50, 51, 67, 68, 69, 78, 103, 104, 106, 109, 133, 190, 226, 266
Mogfna, Anna 109
Morgenthau, Henry 85, 127, 130, 132, 235, 289
Morrissey, Muriel Earhart 86, 127, 215, 249, 256
Myers, Robert xiv, 47, 48, 49, 50, 71, 75, 76, 77, 79, 161, 167, 187, 192, 193, 194, 195, 225, 299, 300

N
Naftel, Saint John 183
Nauticos 145
New Britain Theory 183
Nikumaroro Island 51, 147, 148, 254
Nimitz, Chester 15, 245, 250
Noonan, Frederick 4
Norfolk crash 2
NOURMAHAL, MV 101, 170, 171, 232, 233

O
O'Hare, RM3, C. 53
Oakland Airport 47, 71, 77, 79, 187, 195, 196
Osborne, Carol 127, 164, 254, 256

P
Pacific Islands Monthly 107, 229, 233
Panay, USS 179
Pellegreno, Ann xiii, 194, 221, 249, 251, 252
Philatelic covers 221
Pierson, Carl 61
Popular Aviation Magazine 176
Prymak, Bill 255
Putnam, George P. 34, 37, 87, 176, 212, 213, 215, 216, 219, 220, 225, 226, 228, 229, 232, 233, 234, 236, 238, 239, 247, 256, 257
Putnam, Margaret "Peg" 172, 173
Putnam papers 172, 173, 207

R
Rafford, Paul 50, 82
Reineck, Rollin 51, 100, 149, 168,

189, 192, 195, 257
Reuther, Ron xiii, 206, 214, 256
Riley, John P., Jr. 58
Rokar, Wilbur (See "Rothar, Wilbur") 35
Roosevelt, Franklin, ("FDR") 7, 127
Rothar, Wilbur 22, 26, 29, 32, 33, 34, 36, 37, 39, 117, 190, 205
Royer, Lloyd 44, 83, 84

S
Saipan 9, 11, 13, 14, 15, 50, 86, 88, 92, 93, 94, 95, 96, 97, 107, 108, 109, 134, 135, 136, 137, 139, 140, 156, 159, 184, 190, 191, 195, 196, 198, 199, 201, 212, 213, 226, 227, 228, 236, 237, 238, 247, 248, 249, 251
Samoan Clipper 179, 232
Scheider, Malvina 127, 129
Security Risk Agency (SRA) 201
Sims, Eugene 137
Stultz, Wilmer 1
Sutter, W. L. 53
Swindell, Tod xiii, 27, 152, 153, 192, 203, 222, 250, 253, 298

T
Thompson, RMC G.E. 53
TIGHAR xiii, 51, 86, 145, 147, 148, 182, 254, 303
Tinian 88, 139, 184, 248
True Detective Magazine 117
Truk 9, 12, 74, 84, 107, 134, 158, 159, 180, 189

V
Van Dusen, William 28, 29, 31, 32, 194, 195, 205

W
Weihsien Internment Camp 97
Williamson & Associates 144, 145

About the Author

David Bowman is a Vietnam veteran who retired from the naval reserve in 1991 after 25 years service, which included ten years in Naval Reserve Intelligence. His interest in Amelia Earhart started in 1967 with the publication of THE SEARCH FOR AMELIA EARHART by Fred Goerner, a seminal work on the subject.

In 2002, he retired from a 31 year civilian career with the State of Washington, getting involved with LEGERDEMAIN the following year. Shortly afterward, he had the good fortune to meet David Bellarts as a result of an Internet contact. Mr. Bellarts is the son of Chief Radioman Leo G. Bellarts, who monitored Amelia Earhart's radio transmissions on the U.S. Coast Guard Cutter ITASCA in July 1937.

The author holds two degrees, a BA in Business Administration (Accounting), and a BA in General Studies (Archaeology), the second acquired after three years of night studies at the University of Washington. LEGERDEMAIN is the author's first book.

Printed in the United States
29169LVS00001B/146